SOTHIS PRESS

Sothis Press aims to completely reimagine the public domain book with an emphasis on universal access to captivatingly designed print editions. We believe that technology will enable every single book ever published to be perpetually accessible in beautiful print form. Sothis is beauty in ubiquity.

www.sothis.press
1 Waterhouse Square, London, ECIN 2ST

Practical Problems for Vehicle Draftsmen and Mechanics

Raymond Bernard Birge

SOTHIS PRESS

ISBN 978-1-0365-2254-4

This edition is published by
Sothis Press © 2023. The company
is registered in England and Wales.
The Company Number is 14900389.

R. B. BIRGE and HUGH M. SARGENT, Authors

PRACTICAL PROBLEMS

FOR

VEHICLE DRAFTSMEN AND MECHANICS

WARE BROS. COMPANY, Publishers
1010 Arch Street, Philadelphia, Pa.

CONTENTS

	PAGE
Introduction	4
Geometrical Terms and Definitions	5
To Divide a Line in Any Number of Equal Parts	10
To Lay Out True Sweeps or Curves	11
To Find the Radius of an Arc	13
To Lay Out Ovals	14
The Application of the Proportional Triangle for Laying Out Twisted or Winding Surfaces	16
Construction of Joints	18
Laying Out a Proportional Corner, Using Proportional Triangles	22
Laying Out a Proportional Corner, Using Parallel Lines and Intersections	24
Laying Out Seat Panels, the Dihedral Angle and Miters of a Wagon Seat	26
The Framing of Bodies With Contracted and Flared Sides	28
The Proportional Corner, Horizontal Sections and Corner Strainers	30
Laying Out Twisted or Winding Surfaces	34
Laying Out a Belt Line and Moulding	36
Pricking Off Corner Pillar on Twisted Surfaces	39
Laying Out a Coupé Pillar	42
Framing Up a Door	45
The Construction of Glass Frames	48
Laying Out a Wheel House and Rear Mud Guard	51
Laying Out a Pattern of Panel for a Metal Seat	54
Laying Out a Pattern for a Shroud Panel	57
Laying Out Patterns for a Front Mud Guard and Splasher	60
Perspective Drawing of Vehicles	63
Coloring Carriage and Automobile Drawings	66
Working Drafts	68-80

INTRODUCTION

THIS work is the outgrowth of practical experience in the class room. The authors, while intimately connected with the vehicle industry, in 1909 conducted an evening school in carriage drafting, the pupils of which, about twenty-five in number, were all employes of one large establishment. The results of the first year were so encouraging, and so much benefit was experienced by the students that the work was continued through another year with an equally satisfactory outcome.

As the course of instruction proceeded, many problems came up for discussion and solution in relation to the highly technical work of the draftsman and body maker. These problems were exactly such as might be presented any day in actual practice in any vehicle building factory in which the student finds himself employed. Notes were made of all these practical problems, and their solutions, and these have furnished the basis or foundation for the present volume.

Besides the questions that had been answered and elucidated in the class room, the authors have incorporated in this book a number of very useful original problems, and the whole is now offered to all those ambitious mechanics in the vehicle trade who are seeking to improve their technical education in order to advance themselves in the industry. It was altogether with the object of aiding aspiring mechanics that this book was compiled and placed on the market at a nominal cost.

Some knowledge of elementary plane geometry is indispensably required in order to understand the principles upon which the art of the vehicle draftsman rests. For the benefit of those who have not had the advantage of this preliminary training the subject has been introduced in the first section of the book. A careful perusal of this section will give the student an insight into the principles of plane geometry, and enable him to more clearly understand the technical terms necessarily employed later on. These principles are in fact very simple and easy to understand, but because they are so important and so constantly referred to in practical drafting, it is necessary for the student to have a clear working knowledge of them.

It is the purpose of this book to teach not only the practical applications of each problem, but also to convey a familiarity with the principles underlying each problem. Therefore, in preparing each illustration, those parts have been selected which are as simple as the case may permit, although they are not always in the best proportion. In a number of instances it was necessary to exaggerate the relations of the lines in order to explain the problem to the best advantage.

Great care has been taken to have the drawings and text accurate as the authors could make them, but as most of the work was done by busy men in their evening hours, it is possible that a few errors may have escaped their notice. Should any one discover discrepancies or mistakes of any description in the book the authors and publishers will consider it a favor if the information is brought to their attention.

There has been an insistent demand for a book of this character from the actual workers in the vehicle trades, and if this book meets that demand, or serves in any manner toward elevating the standard of technical education among the vehicle mechanics for whose benefit it was compiled, the authors will feel more than gratified with such a reward for their labor.

R. B. BIRGE.
HUGH M. SARGENT.

Geometrical Terms and Definitions

A LINE is that which has length merely, and may be straight or curved.

A STRAIGHT LINE, or, as it is sometimes called, a right line, is the shortest line that can be drawn between two given points. Straight lines are generally designated by letters or figures at their extremities, as A B, Fig. 1.

A CURVED LINE is one which changes its direction at every point, or one of which no portion, however small, is straight. It is therefore longer than a straight line connecting the same points. Curved lines are designated by letters or figures at their extremities and at intermediate points, as A B C or D E F, Fig. 2.

PARALLEL LINES are those which have no inclination to each other, being everywhere equidistant. A B and A′ B′ in Fig. 3 are parallel straight lines, and can never meet, though produced to infinity. C D and C′ D′ are parallel curved lines, being arcs of circles which have a common center.

HORIZONTAL LINES are lines parallel to the horizon, or level. A Horizontal Line in a drawing is indicated by a line drawn from left to right across the paper, as A B in Fig. 4.

VERTICAL LINES are lines parallel to a plumb line suspended freely in a still atmosphere. A Vertical Line in a drawing is represented by a line drawn up and down the paper, or at right angles to a horizontal line, as E C in Fig. 4.

INCLINED OR OBLIQUE LINES occupy an intermediate between horizontal and vertical lines, as C D, Fig. 4. Two lines which converge toward each other and which, if produced, would meet or intersect, are said to incline to each other.

PERPENDICULAR LINES.—Lines are perpendicular to each other when the angles on either side of the point of meeting are equal. Vertical and horizontal lines are always perpendicular to each other, but perpendicular lines are not always vertical and horizontal, but may be at any inclination to the horizon, provided that the angles on either side of the point of intersection are equal. In Fig. 5, C F, D H and E G are said to be perpendicular to A B. Also, in Fig. 6, C D and E F are perpendicular to A B. Lines perpendicular to the same line are parallel to each other, as C F and D H, Fig. 5, which are perpendicular to A B.

An ANGLE is the opening between two straight lines which meet one another. An angle is commonly designated by three letters, the letter designating the point in which the straight lines containing the angle meet being between the other two letters, as the angle E C D, Fig. 4.

A RIGHT ANGLE.—When a straight line meets another straight line so as to make the adjacent angles equal to each other, each angle is a right angle, and the straight lines are said to be perpendicular to each other. (See C B E or C B D, Fig. 7.)

An ACUTE ANGLE is an angle less than a right angle, as A B D or A B C, Fig. 7.

An OBTUSE ANGLE is an angle greater than a right angle, as A B E, Fig. 7.

GEOMETRICAL TERMS AND DEFINITIONS—Continued

STRAIGHT-SIDED FIGURES.

A **SURFACE** is that which has length and breadth without thickness.

A **PLANE** is a surface such that if any two of its points be joined by a straight line, such line will be wholly in the surface. Every surface which is not a plane surface, or composed of plane surfaces, is a curved surface.

A **SINGLE CURVED SURFACE** is one in which only certain points may be joined by straight lines which shall lie wholly in its surface. The rounded surface of a cylinder or cone is a single curved surface.

A **DOUBLE CURVED SURFACE** is one in which no two points can be joined by a straight line lying wholly in its surface. The surface of spheres, for example, is a double curved surface.

A **HYPOTHENUSE** is the longest side in a right-angled triangle, or the side opposite the right angle. A C, Fig. 8.

The **APEX** of a triangle is its upper extremity, as B, Fig. 9. It is also called vertex.

The **BASE** of a triangle is the line at the bottom. B C and A C, Figs. 8 and 9.

The **SIDES** of a triangle are the including lines. A C, A B and B C, Figs. 8 and 9.

The **VERTEX** is the point in any figure opposite to and farthest from the base. The vertex of an angle is the point in which the sides of the angle meet. B, Fig. 9.

The **ALTITUDE** of a triangle is the length of a perpendicular let fall from its vertex to its base, as B D, Fig. 9.

A **QUADRILATERAL** figure is a surface bounded by four straight lines. There are three kinds of Quadrilaterals: The Trapezium, the Trapezoid, and the Parallelogram.

CIRCLES AND THEIR PROPERTIES

A **CIRCLE** is a plane figure bounded by a curved line, everywhere equidistant from its center. (Fig. 10.) (See also Circumference.)

The **CIRCUMFERENCE** of a circle is the boundary line of the figure. (Fig. 10.)

The **CENTER** of a circle is a point within the circumference equally distant from every point in its circumference, as A, Fig. 10.

The **RADIUS** of a circle is a line drawn from the center to any point in the circumference, as A B, Fig. 10, that is, half the diameter. The plural of radius is radii.

The **DIAMETER** of a circle is any straight line drawn through the center to opposite points of the circumference, as C D, Fig. 10.

A **SEMICIRCLE** is the half of a circle, and is bounded by half the circumference and a diameter. (Fig. 11.)

A **SEGMENT** of a circle is any part of its surface cut off by a straight line, as A E B and C F D, Fig. 12.

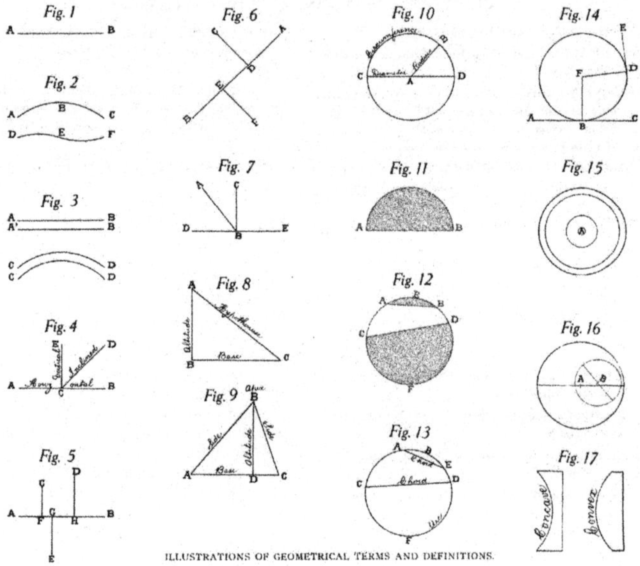

ILLUSTRATIONS OF GEOMETRICAL TERMS AND DEFINITIONS.

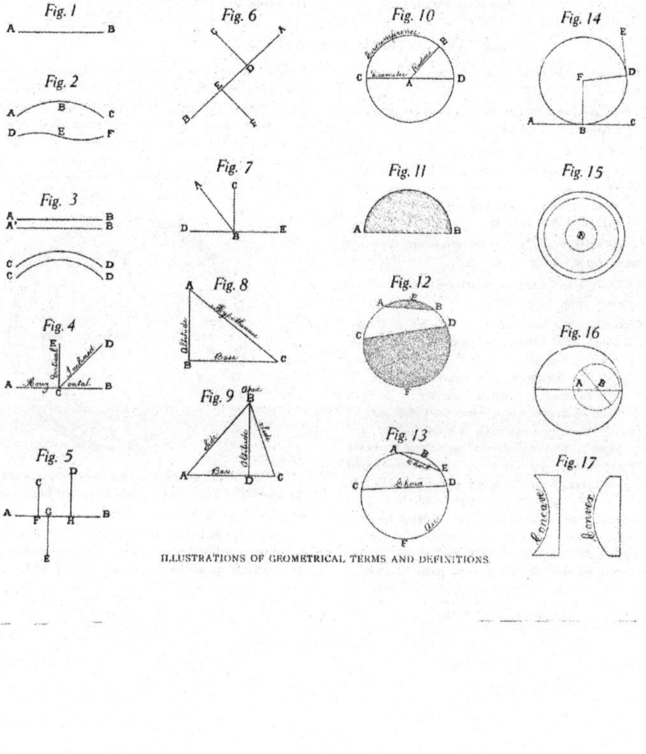

ILLUSTRATIONS OF GEOMETRICAL TERMS AND DEFINITIONS.

GEOMETRICAL TERMS AND DEFINITIONS—Continued

An ARC of a circle is any part of the circumference, as A B E and C F D, Fig. 13.

A CHORD is a straight line joining the extremities of an arc, as A E and C D, Fig. 13.

A TANGENT to a circle or other curve is a straight line which touches it at only one point, as E D and A C, Fig. 14. Every tangent to a circle is perpendicular to the radius drawn to the point of tangency. Thus E D is perpendicular to F D, and A C to F B.

CONCENTRIC circles are those which are described about the same center. (Fig. 15.)

ECCENTRIC circles are those which are described about different centers. (Fig. 16.)

A DEGREE.—The circumference of a circle is considered as divided into 360 equal parts, called degrees (marked °). Each degree is divided into 60 minutes (marked ′); and each minute into 60 seconds (marked ″). Thus if the circle be large or small the number of divisions is always the same, a degree being equal to 1/360 part of the whole circumference; the semicircle is equal to 180°, and the quadrant to 90°. The radii drawn from the center of a circle to the extremities of a quadrant are always at right angles with each other; a right angle is therefore called an angle of 90° (A E B, Fig. 18). If a right angle be bisected by a straight line, it divides the arc of the quadrant also into two equal parts, each being equal to one-eighth of the whole circumference, or 45° (A E F and F E B, Fig. 18); if the right angles were divided into three equal parts by straight lines, it would divide the arc into three equal parts, each containing 30° (A E G, G E H, H E B, Fig. 18). Thus the degrees of the circle are used to measure angles, therefore by an angle of any number of degrees, it is understood that if a circle with any length of radius be struck with one foot of the compasses in its vertex, the sides of the angle will intercept a portion of the circle equal to the number of degrees given. Thus the angle A E H, Fig. 18, is an angle of 60°.

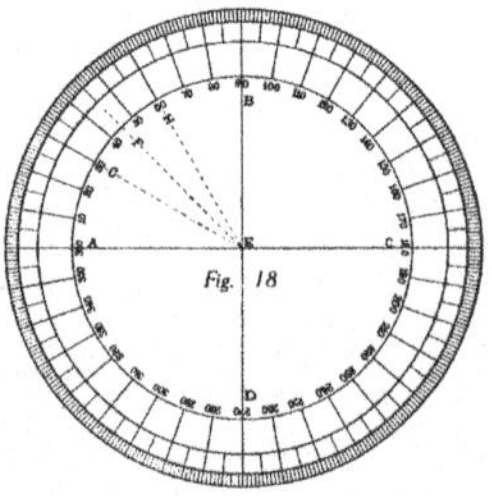

Fig. 18

CONCAVE means hollowed or curved inward, said of the interior of an arched surface or curved line in opposition to convex. (Fig. 17.)

A CONVEX surface is one that is curved outward, that is regularly protuberant or bulging, when viewed from without. The opposite of convex is concave. (Fig. 17.)

To Space Off Equally a Number of Parts on a Line, the Length of Which is Given

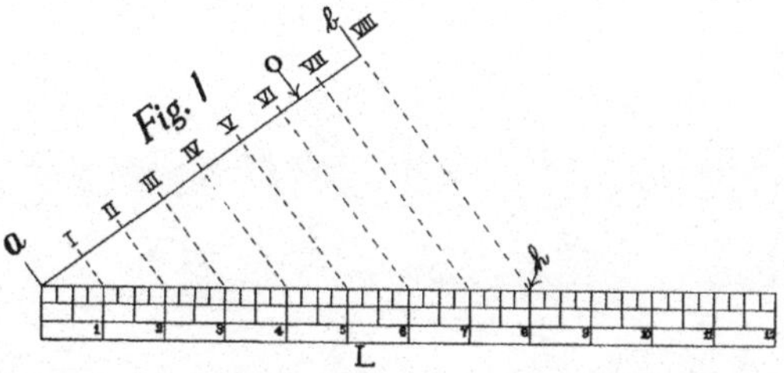

SAY that one wishes to divide a line 6½ inches long into eight equal spaces. The easiest and quickest method is as follows:

Lay off line O, Fig. 1. Draw a perpendicular line from b, and place the scale L from a until 8 inches strike the perpendicular line drawn from line O at h. Mark each inch from the scale on to the drawing, and project on to line O, establishing points I, II, III, IV, V, VI and VII, which are equally spaced on line O.

The Laying Out of True Sweeps or Curves when the Length and Amount of Sweep or Curve is Given

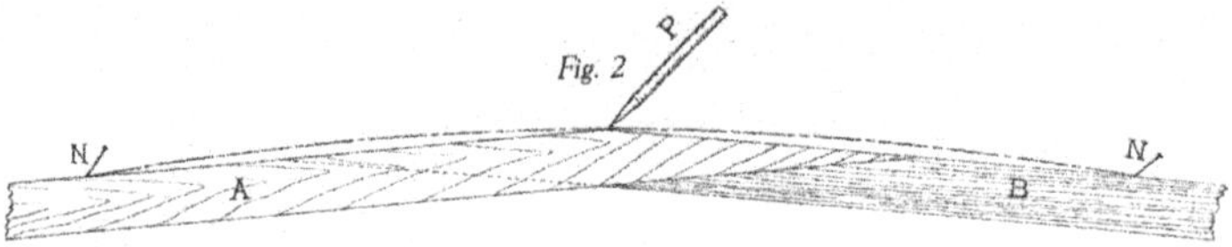

Fig. 2

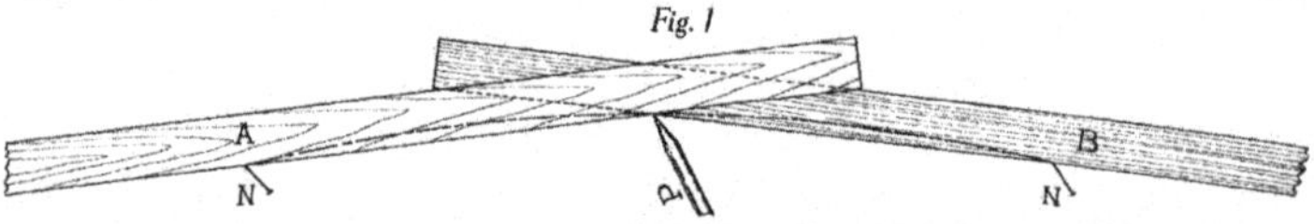

Fig. 1

FOR laying out true sweeps or curves where it is impractical to strike the radius with the compasses or trammel points, we illustrate the following example, and although the principle has long been in use, it is unknown to a great many mechanics.

For the example let us say that we desire a true sweep of 4 inches in a length of 40 inches. Lay out on a straight line two points 40 inches apart, and at each point place a nail as illustrated in Fig. 1. Half way between the two points lay out on a perpendicular line a point 4 inches from the base

THE LAYING OUT OF TRUE SWEEPS OR CURVES WHEN THE LENGTH AND AMOUNT OF SWEEP OR CURVE IS GIVEN—Continued

line. Place at this point another nail. Next take two sticks, each a little longer than the length of the sweep, and secure them together, resting on the nails and forming a triangle, as shown. A and B in Fig. 1 represent the sticks, and N represents the nails.

After the sticks are placed tight up against the nails and fastened securely so that they cannot change their angle, remove the top nail, and notch out the sticks at this point so that a lead pencil P may be placed therein.

To obtain the sweep, work the sticks back and forth between the nails N-N in Fig. 1, being sure to have the sticks bear against the nails. In this way a sweep of any length may be obtained, and this system will be found very convenient in cases where the radius is too great to be reached by the compasses or trammel points.

It is usually customary to form the sticks as shown in Fig. 1, and place the pencil on the inside of the sticks as shown, but in some cases it is desirable to construct the sweep as shown in Fig. 2, with the outside of the sticks bearing against the edge of the nails, and the pencil being placed in the notch on the outside of the sticks. Although this is sometimes desirable, we believe it will be found more convenient to lay a sweep out as illustrated in Fig. 1, inasmuch as one need not be particular in fastening the sticks together, except that they must be fastened securely so that they will not change their position in relation to the nails.

To Find the Radius of an Arc

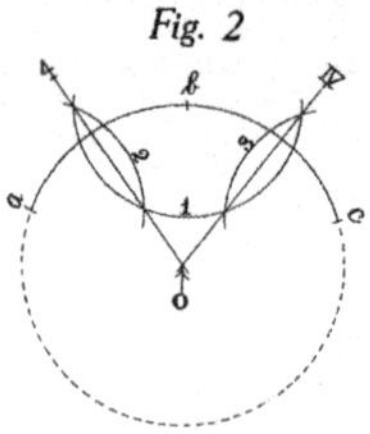

AN ARC FOR WHICH WE WISH TO FIND THE CENTER OR THE RADIUS.

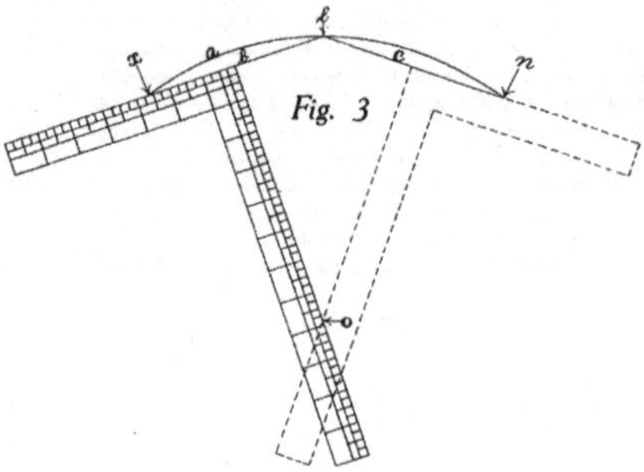

ANOTHER METHOD OF FINDING THE RADIUS.

REFER to a-c in Fig. 2, an arc for which we wish to find the center or the radius that described same. Take point b, which is midway between a and c, and describe arc 1 therefrom, the radius of which is greater than half the distance from a to b, or from b to c. With the same radius centered at a and c, describe arcs 2 and 3, passing through arc 1 described from b. Draw the lines 4 and IV through intersection of arcs, and where these lines intersect at O we will have established the center from which the large arc a-c was described.

Fig. 3 illustrates another method for finding the radius of an arc by the use of a square. Find the center of the arc x-n which is at l. Draw straight lines between points x, l and n. Find the center of line l-n which is at c, and with the square placed on line l-n, draw a perpendicular line from c. Then with the square placed on line x-l, draw another line perpendicularly from the center of line x-l at b until line intersects that drawn from c at O, which is the center of the arc x-n.

Laying Out Ovals

Fig. 1

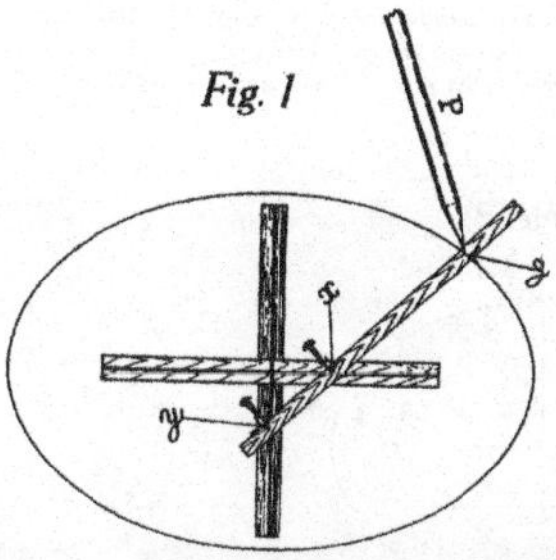

ONE SYSTEM FOR FINDING AN OVAL.

THERE are several systems for laying out a true oval, and in this article we will explain and illustrate three of the most popular.

The first system is illustrated in Figure 1. A square frame is made in the shape of a cross, and is grooved out as shown to accommodate two adjustable pins or studs secured to a stick, as shown by y-x. Decide upon the length and width of the oval desired, say 18 inches long and 9 inches wide. Lay out from the pencil point on the stick one-half the width of the oval, which is 4½ inches, from z to x. Locate one of the adjustable pins at x. With one-half the length of the oval, which would be 9 inches, lay same off on the stick from z to y. Set the adjusted pins into the grooves in the cross frame and work the same up and down, and back and forward on the frame, having the pencil point bearing on the paper and describing a true oval around the cross frame.

Figure 2 illustrates another method for laying out an oval, and when there are no instruments at hand, same will be found very convenient. Lay out the horizontal and vertical center lines of the oval, and on same lay out the length and width desired. Take one-half the length of the oval from x to a and lay this distance out from c until same strikes the horizontal

LAYING OUT OVALS—Continued

center line of the oval at e and h, at which points locate two pins or tacks, as illustrated. At point c locate another pin or tack, and around these three pins or tacks tie a piece of string taut, forming a triangle e-h-c. Remove the pin or tack at c, and put the point of a pencil in its place. Keeping the

Fig. 2

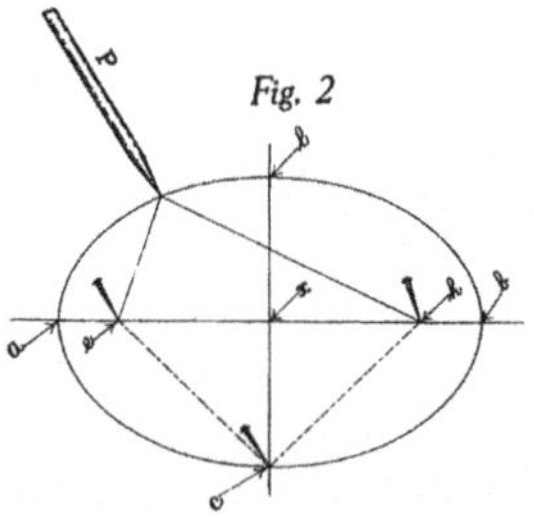

ANOTHER SYSTEM FOR FINDING AN OVAL.

string taut, with the pencil back of same, work the pencil P around the pins, which operation will produce a true oval.

Fig. 3 shows another handy system for laying out a true oval. Draw the horizontal and vertical center lines of the oval, and determine the length and width. From O, the center of the oval, draw a circle B, the diameter equal to the width of the oval. Draw another circle A from the center O, the diameter of which will be equal to the length of the oval. Space off a number of points 1, 2 and 3 on the outer circle A, and connect with the center of the oval at O, and where lines drawn from 1, 2 and 3 intersect the inner circle B, draw

Fig. 3

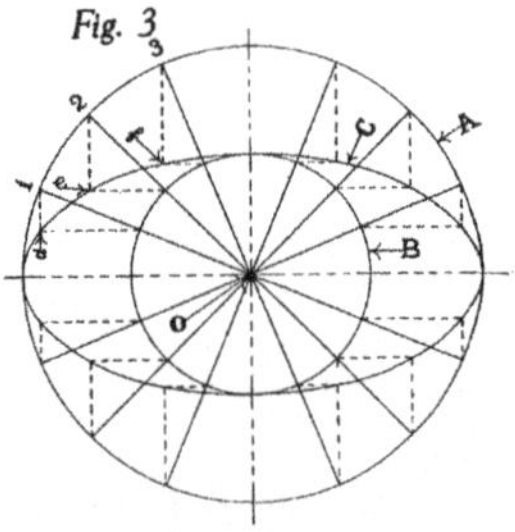

LAYING OUT A TRUE OVAL.

horizontal lines, outward, and from points 1, 2 and 3, on the outer circle A, drop vertical projections until same strike the horizontal lines drawn from the inner circle B at d, e and f. The same operations apply for obtaining the remaining points required, and after same are found it is necessary to connect them with graceful sweeps which will produce the true oval C.

The Application of the Proportional Triangle for Laying Out Twisted or Winding Surfaces

THE proportional triangle will be used to a great extent in many of the following problems for laying out proportional turn-unders of twisted or winding surfaces, and we believe that these advance problems will be found invaluable in preparing the student for some of the more difficult ones to be taken up in some of the following problems.

Let us first explain Fig. 1. The turn-under sweep B-Z is an assumed line, and from this line, providing we know the over-all amount of turn-under, we may work proportional turn-under sweeps for any part of the body. B-Z being the assumed turn-under sweep, we desire to find out the correct proportional turn-under sweep for another part of the body which with the over-all amount of turn-under is equal to the distance from A to C on line O-O, Fig. 1.

Let the vertical line Z-X be drawn through the horizontal line O-O, and lay out a triangle BCX below line O-O. Point B is fixed by the assumed turn-under sweep. Point C is fixed by the vertical line Z-X as drawn from the top of the body and passing through base line O-O, and, in completing the triangle, the only point we have to determine is X, and this may be located arbitrarily on the vertical line Z-X and to suit the conditions. We have point A also fixed by the over-all amount of turn-under sweep which we wish to proportion, so that we will lay out another triangle ACX, X being the common apex for both triangles.

Now we have laid out a triangle for each turn-under, and it is by means of these triangles and the assumed turn-under sweep that we will develop the greater turn-under sweep A-Z. Passing through the assumed turn-under sweep B-Z and the vertical line Z-X, lay out a number of lines I, II and III at will above the horizontal line O-O. Where each one of these lines passes through the assumed turn-under sweep B-Z, take I for instance, drop a vertical line until we strike the hypothenuse for the proportional triangle of the assumed turn-under sweep. From this point draw a horizontal line passing through the triangles and number it 1. Where this line strikes the hypothenuse of the larger proportional triangle, erect a perpendicular line until line I is reached, thus establishing a point through which the correct proportional turn-under line A-Z must pass. Continue this operation with the remainder of the points, and connect with a graceful sweep, and we will have obtained a correct proportional turn-under, line A-Z, to the assumed turn-under, B-Z.

In Fig. 2 we illustrate another example which is equally important as that of Fig. 1. When a body is constructed with untwisted sides as far as the seat bottom, and in this illustration, Fig. 2, we will consider that the seat frame or the commencement of the twist is from line IV. B-Z is the assumed turn-under sweep. A-Y is the amount of turn-under at another part of the body for which we wish to proportion the turn-under sweep, having the same twist from the under

THE APPLICATION OF THE PROPORTIONAL TRIANGLE FOR LAYING OUT TWISTED OR WINDING SURFACES—Continued

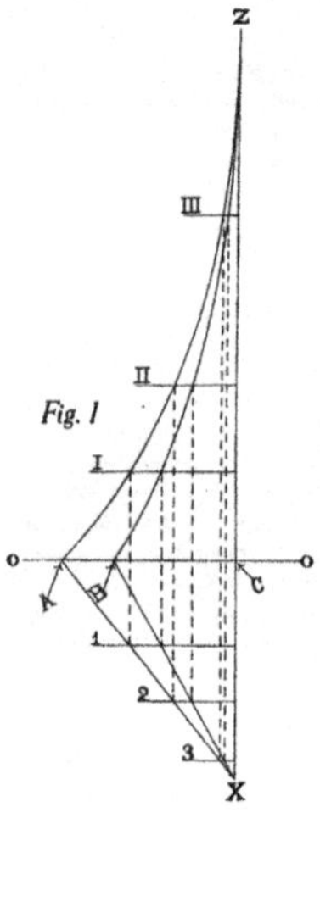

Fig. 1

side of the seat or line IV. Draw the straight line Z-X as a continuation of the body line before it commences to turn under at IV. It is on line Z-X that we locate the common apex X of our proportional triangles, which, as in Fig. 1, is determined at will. Lay out the proportional triangles BCX and ACX for the assumed turn-under sweep and the desired turn-under sweep, respectively. Space off a number of lines I, II, III and IV passing through the assumed turn-under sweep. Wherever these lines cut the assumed turn-under sweep, line B-Z, draw lines down on to the proportional triangle parallel with the outside line Z-X. Where these parallel lines strike line B-X of the triangle BXC, draw horizontal lines 1, 2 and 3 passing through the line A-X of the larger proportional triangle ACX. Project these points up and parallel to the outside line Z-X until lines I, II and III are intersected, thus establishing points through which the desired turn-under sweep will pass.

The only difference between this problem and that in Fig. 1 is that, instead of the usual vertical projections from the assumed turn-under sweep to the proportional triangles, the projections are parallel with the outside flared line Z-X. Fig. 2.

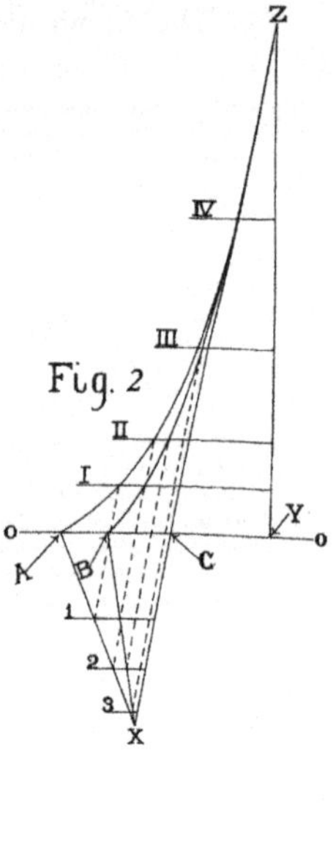

Fig. 2

Construction of Joints

TO GO INTO this subject as we would like would require more space in this book than can be spared; therefore we will explain briefly the ordinary construction of the various joints used in body-making, and we believe that the illustrations will be found of great value to the beginner, and should be an incentive for further study in this very important subject.

HALVED OR LAPPED JOINTS

A half-lap joint varies in shape and arrangement with the importance of the pieces united. A full lap is used when one piece is let into the other to its full thickness; a half lap when each piece united is cut away, whatever the shape of the lap may be. The common expression, "half-lapped," does not mean that each piece united must be cut half away.

Fig. 1 shows the framing up of a toe bracket, each piece being half-lapped into the other. In the sill and angle piece there is indicated a rabbet for the floor board, as shown by sections a and b. In Fig. 2 is shown a batten or strainer framed into the sill with a feather-edge lap joint. As this batten or strainer has little work to do other than that of holding the panel in shape and supporting it, this kind of a joint is preferable to one more complicated. Fig. 3 illustrates the framing of a roof bow into the head rail of an enclosed body using a half-lapped joint. Fig. 4 illustrates a section of an under body with the post half-lapped into the sill and top rail. As an illustration of a full-lap joint we show the floor board let into a rabbet in the sill.

MORTISE AND TENON JOINT

Fig. 5 shows the two pieces to be united which form this joint. Note that they are both of the same thickness, and A is cut away equal on both sides for the tenon. The projecting part c on A, Fig. 5, is the tenon, and d in B is the mortise. e is the shoulder of the tenon, which is equal on both sides. f is usually one-third of the thickness of A and B.

BARE-FACED TENON AND MORTISE

In Fig. 6 we illustrate a bare-faced tenon and mortise joint. This means that the face of the tenon, or both faces, are exposed. The bare-faced tenon and mortise is used when two pieces are joined endways in the same direction, or endways in different directions, as shown in Fig. 6.

STUMP TENON.

Fig. 7 shows a mortise, tenon and stump joint, and is frequently used in framing up parts where certain joints should not be exposed as they are in Fig. 6. The stump reinforces and prevents twisting.

OPEN MORTISE, TENON AND MITER JOINT

This joint is illustrated in Fig. 8, and is used in framing up parts where it is undesirable to have the joints exposed.

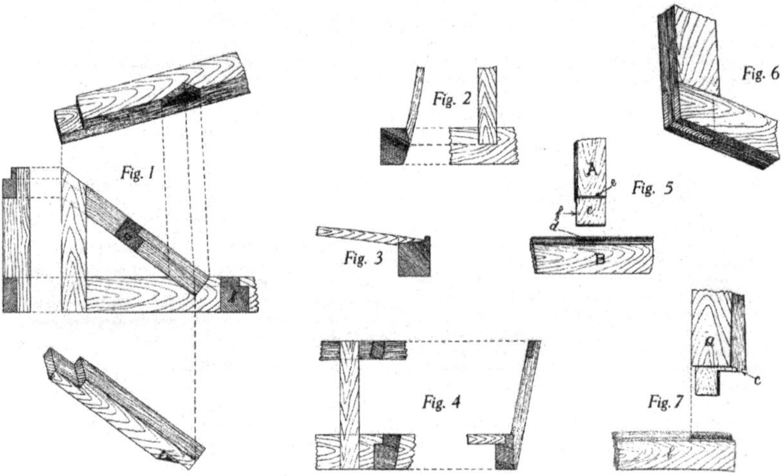

ILLUSTRATING THE CONSTRUCTION OF JOINTS.

CONSTRUCTION OF JOINTS—Continued

DOUBLE TENON AND MORTISE

The double tenon and mortise joint is illustrated in Fig. 9, and is used for framing the fence bar a into the door pillar b. This, of course, makes a stronger joint than with a single tenon, but in most cases a single mortise and tenon joint is sufficient. A rabbet is also illustrated in Fig. 9, the lock board c being let into the pillar b.

FALSE TENON

Fig. 10 shows a false tenon for joining a frame, as illustrated. In this case a false tenon is used on account of its simplicity, but is not as satisfactory as a bare-faced tenon and mortise joint. In framing up parts with a false tenon, always be particular to have the grain of the tenon cross that of the grain of the parts united.

MITER JOINT.

Fig. 11 shows a simple 45-degree miter joint. This is self-explanatory.

STUMP TENON AND LAP JOINT.

Fig. 12 illustrates a stump tenon and lap joint, which is much stronger than the half-lap, and is frequently used in framing up side rails and pillars.

BASTARD TENON

This joint is shown in Fig. 13 in the framing of a post into a sill and where exposed joints are undesirable. It will be seen that the half-lap joint would show two long joints, while a mortise and tenon joint would be impractical inasmuch as the tenon would be too close to the outer surface.

Fig. 14 illustrates dovetail joints. Figs. 15 and 16 show ordinary splice joints. Fig. 17 illustrates a method for securing metal panels to body framing, and by this method we do away with exposed screw heads and nails, the metal panel being turned around a 1/4" x 1" flat iron which is let into a rabbet flush with the standing pillar. This is secured by means of machine screws from the inside of the post which are covered up by the trimming.

Fig. 18 illustrates a method for securing the upper panels, above the belt, of a limousine body. Note that this upper panel is offset from the lower panel, and that the belt rail is rabbeted out to receive an iron plate and around which is formed the upper metal panel. The lower panel also comes up back of this iron, forming a tight joint, the iron being secured by means of machine screws from the inside of the belt rail, as illustrated. This upper offset panel is generally made of wood, but this construction eliminates the troubles incident to wood panel construction, over which the metal panel has the advantage.

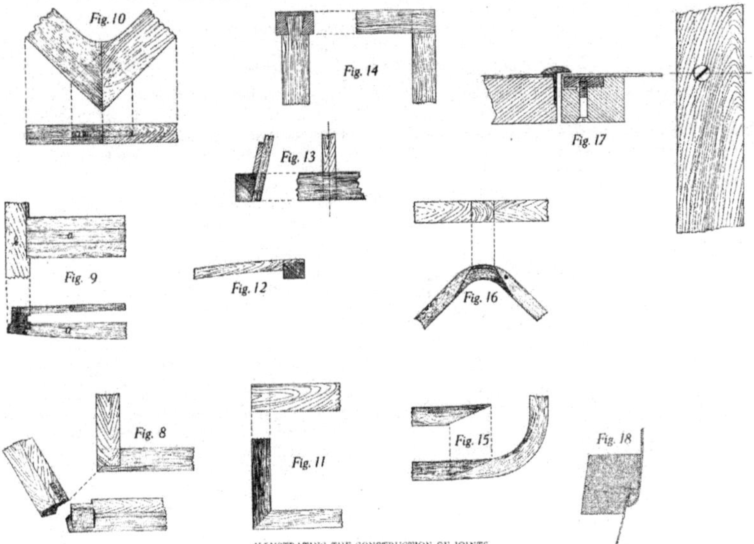

ILLUSTRATING THE CONSTRUCTION OF JOINTS.

Laying Out a Proportional Corner by the Use of Proportional Triangles

FIG. 1 is the profile or side view of an automobile seat. Fig. 2 is one-half the bottom view, and Fig. 3 is the side bevel or flare of the seat. Lay out the horizontal line EC in Fig. 1, drawn from the extreme height of the seat. This line is partly imaginary, but is necessary in laying out the arm rail in Fig. 2.

To proportion the corner in Fig. 2 it is necessary to assume either the top line around the seat, or the bottom line. In this case we take the top line EC, Fig. 2.

On Fig. 1 lay out the proportional triangle FOB. BF is the amount of back bevel squared down to the bottom of the seat from C. The apex O should be located on the horizontal line EC at the beginning of the proportioning.

Next lay out the proportional triangle EQA, Fig. 2. AE is the amount of side bevel or flare taken from Fig. 3, a to c. Q is the apex of the triangle, and should be located on the center line XC. The line AQ of the proportional triangle, Fig. 2, is parallel to OB of the triangle on Fig. 1.

With these triangles located, lay off arbitrarily on line EC, Fig. 2, points 1, 2, 3, 4, 5, 6, 7 and 8 around the corner. Starting at 1, square point up to the outside line of triangle at j in Fig. 1. Then square point ahead until it strikes inner line of triangle, Fig. 1, at k. From this point drop a perpendicular line on to Fig. 2.

From point 1 on line EC, Fig. 2, lay out horizontal projection striking outside line of proportional triangle, Fig. 2, at l. Drop a perpendicular line on to inner line of triangle

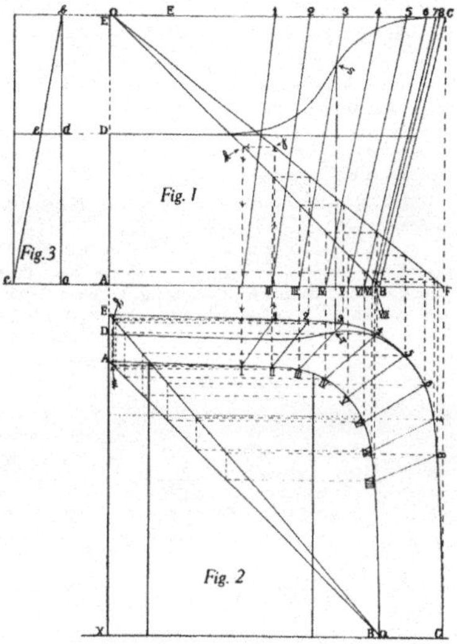

SIDE VIEW, ONE-HALF BOTTOM VIEW AND SIDE BEVEL AND FLARE OF AN AUTOMOBILE SEAT.

LAYING OUT A PROPORTIONAL CORNER BY THE USE OF PROPORTIONAL TRIANGLES—Continued

at m. From m draw a horizontal line back until it intersects vertical line drawn from k, Fig. 1, thus establishing the point I which is proportional to point 1 on line EC, Fig. 2. Continue this operation with 2, 3, 4, 5 and 6, etc., and we will have obtained points on the seat frame through which the bottom line AB in Fig. 2 must pass.

In order to lay out arm rail, 6 to D on Fig. 2, proceed as follows:

Connect points 1, 2, 3, 4, 5 and 6 on line EC, Fig. 2, and corresponding points on line AB with straight lines. Project these same lines and points on to Fig. 1. Wherever the arm rail line cuts these lines, take s for an example, Fig. 1, square this point down until it intersects the corresponding line on Fig. 2 at T. Repeating this operation with all of the other intersections will produce the points on Fig. 2 through which the arm rail line should pass.

D shows the amount of side flare at the front of the arm rail on line ed, Fig. 3. This same method of projection can be applied for obtaining any intermediate section between lines AB and EC, provided the flare of the seat is straight all around. Do not attempt to use this system of projection if the turn-under is either concave or convex, but use the proportional scale as explained and illustrated elsewhere in this book.

Laying Out of Proportional Corners by the Use of Parallel Lines and Intersections

IN THIS problem we illustrate and explain another system for laying out proportional corners of seats, etc. This problem has no connection with the problem taken up previous to this for laying out proportional corners, but is simply another method which is used a great deal, and we leave it to the student to select whichever is most convenient for his use, although we have adopted this system throughout the book, and believe it the handier of the two.

It is first necessary to draw out the seat in the profile as in Fig. 1, and the partly imaginary line a-O drawn horizontally from the extreme height of the seat or body. In Fig. 3, the plan view, assume either the bottom seat line d-D, or the partly imaginary line a-A taken from the horizontal line at the extreme height of the seat in Fig. 1. In this problem we suggest laying out the top line around the seat a-A to begin with, and from this line in Fig. 3, after we have obtained the flare of the side in Fig. 2 and the flare of the back in Fig. 1, and transferred them to Fig. 3, we will obtain the proportional line of the seat bottom and intermediate sections.

On the top line around the seat, a-A in Fig. 3, lay out a number of points, namely, 1, 2, 3, 4 and 5, same being located on the sweep or curve of the corner. With the use of these points we will proceed to obtain points corresponding to these on the bottom line of the seat, Fig. 3, through which the same line of seat will pass. From point 6 on the center line of the seat in Fig. 3 draw a straight line connecting with point 5 on the top line of the seat, and parallel to this line from point VI draw another line extending indefinitely through the seat.

From point A on the partly imaginary line a-A in Fig. 3 draw a straight line, and connect same with point 5 on line a-A, and parallel to this line draw another line from D on the front of the seat at the bottom until same line intersects the line drawn from VI at V.

From each of the remaining points, 1, 2, 3 and 4, on the top line of the seat a-A, Fig. 3, draw lines to points 6 and A on line a-A, and parallel to these draw lines from VI, and from D until lines drawn from D intersect those drawn from VI at I, II, III and IV. This establishes proportionally points through which the bottom line of the seat d-D should pass.

Now that we have the top and bottom lines of the seat, it is essential to prick off the arm rail line in Fig. 3, which is accomplished as follows:

Project from Fig. 3 to Fig. 1 on to their relative positions lines 1-I, 2-II, 3-III, 4-IV and 5-V. Wherever the arm rail line passes through these oblique lines in Fig. 1 drop vertical projections until same strikes corresponding oblique lines in Fig. 3, thus establishing points through which the arm rail line shall pass. Take the amount of flare at the front of the seat and the height of the side in Fig. 2 and transfer from D to B in Fig. 3. Draw graceful line passing through the points already obtained, and same will produce the arm rail line in Fig. 3.

Now, we are supposing that the construction of this seat calls for a belt moulding or rail around the seat at the height of c-C in Fig. 1. We wish to produce this line around the seat in Fig. 3, and to accomplish same we will take points where line c-C in Fig. 1 passes through the oblique lines 1, 2, 3, 4 and 5 and project these points down on the corresponding oblique lines 1, 2, 3, 4 and 5 in Fig. 3. Take the amount of flare at the height of C from Fig. 2, and lay same out in its correct position from D to C on the front line of the seat in Fig. 3. Also take the amount of flare at the height of c from the back of the seat in Fig. 1, and transfer from d to c on the center line of the seat in Fig. 3. Connect these points with the points already obtained on lines 1, 2, 3, 4 and 5 in Fig. 3, and we will have produced line c-C in Fig. 3, corresponding to line c-C in Fig. 1, which is proportional with top and bottom sweeps of the seat in Fig. 3.

Likewise any point on the bottom view of the seat in Fig. 1 can be obtained in the same manner, the system being that of projection.

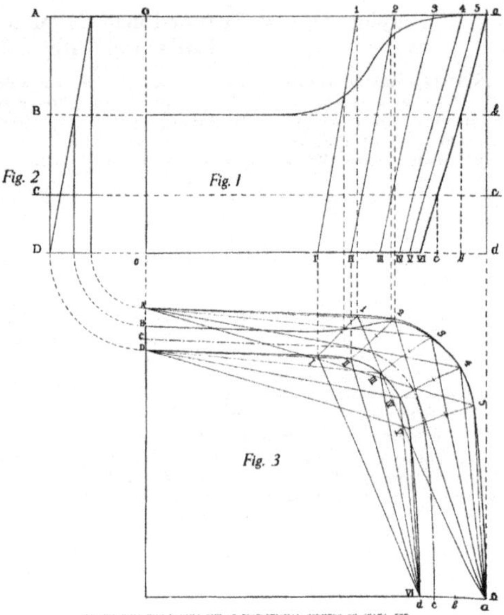

ILLUSTRATING THE LAYING OUT OF PROPORTIONAL CORNERS OF SEATS, ETC.

Laying Out Seat Panels, the Dihedral Angle of a Corner Block, and Miters of a Wagon Seat

LET FIGS. 1 and 2 represent the side elevation and one-half back view, respectively, of a wagon seat. Line bc in Fig. 1 represents the back flare; line bc in Fig. 2 represents the side flare. In order to lay out the exact size of the panels, proceed as follows:

Fixing the compasses at b in Fig. 2, describe an arc from c until it strikes the vertical line at H. Project point from H over on to Fig. 1 until it strikes the vertical line F at C. Connecting b and C, Fig. 1, gives us the exact flare of the rear end of the side panel. Lay out a line DC parallel to the top line of the side panel dc, Fig. 1, until it strikes the vertical projection from d at D. Connect points D and a, Fig. 1, which will give us the front line of the seat panel. Hence the dot and dash lines connecting a, D, C and b, Fig. 1, represent the exact size of the stock for the side panel.

For the full size of the stock for the back panel, set the compasses at b in Fig. 1 and describe an arc from c, striking the vertical line at E. Square this point on to vertical line C, Fig. 2, and connect with b, Fig. 2, as shown by dot and dash line, and carry the top line of the panel parallel to line cx, thus producing the size and shape of the stock for the back panel.

Now that we have the size of the seat panels, it is desirable to determine the dihedral angle or corner block bevel. From point III, which is taken arbitrarily, on the line of the stock, Fig. 1, draw a line perpendicular to the line of the stock bC, Fig. 1, until it strikes the base line at V. From the line of the stock for the back panel, bC in Fig. 2, square a line from II down to the base line at I. Point II, Fig. 2, is a horizontal projection from III, Fig. 1.

The next operation is to draw a line perpendicular with the base from point b in Fig. 2. Take the distance from V, Fig. 2, to b, Fig. 1, and lay same off from b to V on the vertical line in Fig. 2. Take the distance from point V, Fig. 2, to III, Fig. 1, and as a radius describe arc VI from point V on the vertical line b. Next describe an arc from point I, Fig. 1, with a radius equal to the line from I, Fig. 1, to II, Fig. 2, this arc VII intersecting arc VI at Z.

Now the dihedral angle or corner block bevel will be taken from the angle IZV. In order to find the bevel of the mitres for the seat panels, simply take the corner block bevel and bisect same as illustrated in Fig. 4.

For further explanation, the dot and dash lines in Figs. 1 and 2 indicate the shape and size of the seat panels or stock when laid flat, marked by S, Figs. 1 and 2. In Fig. 3 the corner block B is illustrated, and should be planed up with the bevel as taken from IZV.

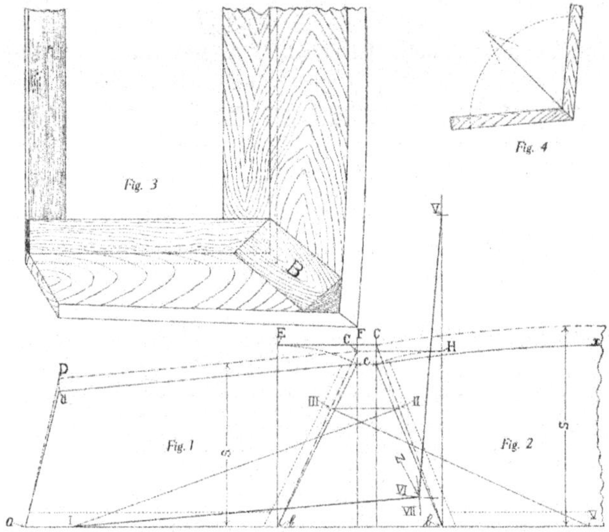

LAYING OUT SEAT PANELS AND FINDING THE DIHEDRAL ANGLE.

The Framing Up of Bodies Having Contracted and Flared Sides

IN THIS article we illustrate two problems, Figs. 1 and 2, which represent portions of seat framing showing an upright post framed into a seat frame and top rail. The object of these lessons is to show how to obtain the bevels of the joints, and the angle or sail that should be given the post in order to make it stand in its correct position when set up with the contraction and flare.

Whenever a seat is constructed with a post standing square or plumb, as in the side view, A, Fig. 1, the post cannot be framed up square with the contracted lines of the seat frame and top rail, but it must be set at a certain angle, so that when same is contracted or flared it will stand square or plumb in the vertical plane. The amount of contraction and flare determines the amount of bevel that should be given to the joints if the post is desired to stand square in the vertical plane A.

To obtain the required bevel of the joints or angle of the post, proceed as follows:

In Fig. 1 lay out the side view A, having the post square or perpendicular with the base line. Determine the amount of side flare desired, and lay out on the end view B, and on the bottom view C show the amount of contraction desired. From point I, the back edge of the post where it intersects the top rail line b-d in C, Fig. 1, draw a line perpendicular with the top rail, line b-d, until it strikes the bottom line of the seat at III. Take the distance from II, where the back edge of the post intersects the bottom edge of the seat frame, to III on the bottom view C, and transfer from II forward to III on the bottom line of the seat in the side view A. Draw a straight line passing through I and III which will establish the angle or sail on which the post must be framed up in order to have it stand square or plumb in the vertical plane A after it is thrown around on the contraction and flare.

Fig. 2 illustrates a similar example, the only difference being that the post is set up on an angle or sail in the vertical plane A. In framing the post into the top rail and seat frame the bevels for the joints must not be taken directly from the side view A, but should be determined by means of the following system:

From the contraction line b-d on the bottom view C, square a line from the intersection of the back edge of the post with the top of the top rail at I until it strikes the bottom of the seat frame a-e at IV. From point I on the bottom view draw a line square across the body, intersecting the bottom of the seat frame line at III. Take the distance from III to IV on line a-e in the bottom view C, and lay same out on the bottom line of the seat in the side view A from the back of the post II forward to III. Connect III and I with a straight line, thus establishing the angle or sail on which the post must be framed up in order to have it conform, after it is swung around on the contraction, to the angle or sail shown in the vertical plane A. This line also establishes the bevel on which the shoulders for the joints should be cut.

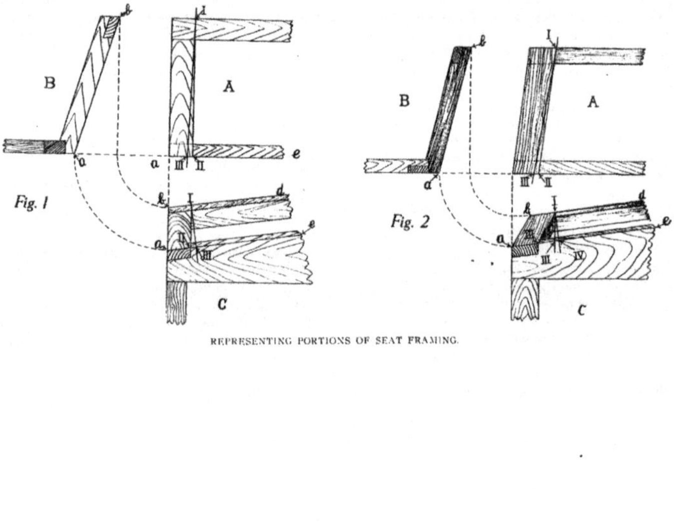

REPRESENTING PORTIONS OF SEAT FRAMING.

Position of Proportional Corner and Laying Out of Horizontal Sections and Corner Strainers on a Torpedo Body

THESE problems are illustrated on the rear of a torpedo body. First let us explain the position of the proportional corner. Referring to page 24, we will find the system desired for obtaining the bottom line around the corner of body, after the top line is assumed.

On page 25 it will be noted that the proportioning was commenced at the front of the seat on the side, and the center of the seat at the rear. The points from which the proportioning begins regulate the size of the corner somewhat at the bottom. This, of course, is a matter of taste. If one desires a large corner at the bottom, he may work the corner as on page 25, but the method we introduce in this article has proven very satisfactory, and we recommend it on bodies of this type. As will be noted, it gives a smaller corner on the bottom. After the side view or profile of the body is determined in Fig. 1, the width from the center line to Z in Fig. 2, and the turn-under, Fig. 3 should be laid out on the drawing.

The first operation is to assume the line Z-Y, Fig. 2, taken at the height of Z-Y, Fig. 1, and Y-Z, Fig. 4, but care should be taken to produce a smooth and graceful line around the corner.

Wherever the top corner sweep forms tangents to the side and back sweeps, as at f and e, Fig. 2, draw horizontal and vertical lines, respectively, until they intersect at H, Fig. 2.

Project point f, Fig. 2, up until it strikes the horizontal line Z-Y, Fig. 1, and from this point draw a straight line parallel to t until it strikes base line X-T at g. Square this point down to the side sweep, Fig. 2 at g, and this gives us the commencing of the proportional corner on the side.

To find out where to commence the proportional corner at the rear, lay out points A and g, Fig. 4, which correspond to A-g, Fig. 2. Connect these points on Fig. 4 with a straight line.

From e on the horizontal line Y-Z, Fig. 4, which is taken from center line to e, Fig. 2, draw a line at any convenient angle, say 45 degrees, until it strikes straight line connecting A and g at j, Fig. 4, from which draw a horizontal line until it strikes center line of rear view at i. Connect i and e, and we will have completed a triangle e-i-j.

Draw another triangle mkl, Fig. 4, with the sides parallel to those of the triangle e-i-j, Fig. 4. At any convenient point, l for example, which is located on straight line passing through A-g in Fig. 4, draw a line parallel to e-j. From point k on center line of rear view, which is a horizontal projection from l, draw a line parallel to i-e, and these two lines intersecting at m form the second triangle, mkl, Fig. 4.

Connect the apexes e and m of the triangles with a straight line, and the point where same cuts the horizontal base line T-X at h, Fig. 4, gives us the point h on Fig. 2, from which draw a horizontal line until it cuts the back top sweep at M.

Now we have on Fig. 2 points A, g and M, h, from which

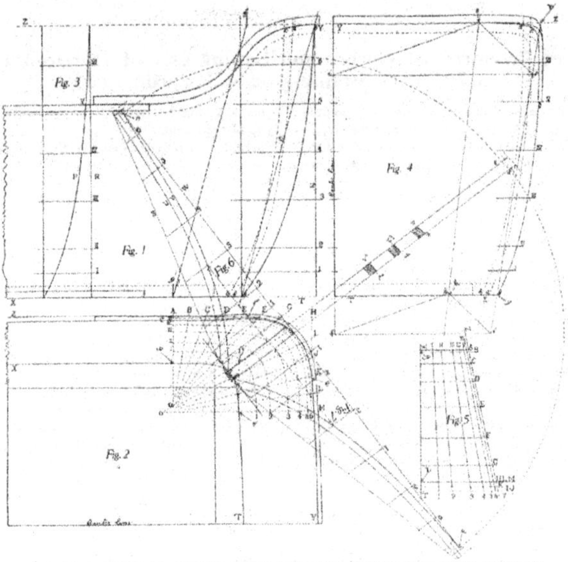

ILLUSTRATING THE PROPORTIONAL CORNER, HORIZONTAL SECTIONS AND CORNER STRAINERS.

POSITION OF PROPORTIONAL CORNER AND LAYING OUT OF HORIZONTAL SECTIONS AND CORNER STRAINERS ON A TORPEDO BODY—Continued

proceed to lay out from the top line the points necessary to obtain the bottom line of the corner, as in problem on page 25. After this is accomplished, continue the bottom side sweep X-g parallel to ZA, Fig. 2. Also lay out the bottom rear sweep from h parallel to M-Y.

Now it is necessary to have sections of the corner at various heights between the top and bottom lines of the seat from which we can make the necessary forms to enable us to obtain the correct shape of the seat or body at a given point accurately.

Draw horizontal and vertical lines from g and h, respectively, until they intersect at U, Fig. 2. Draw a straight line through H and U, and a horizontal line through M and h until they intersect at O, Fig. 2.

Lay off a convenient number of points around the corner between A and M on line Z-Y, Fig. 2, as B, C, D, E, F, G, H, I, J, K and L.

From points between H and M only, Fig. 2, draw straight lines through the body until they converge at O, as from I to O, J to O, K to O, etc.

Square down through A and g a line intersecting line OH at Q, Fig. 2. Connect points B, C, D, E, F, G, between A and H, with point Q.

Determine locations of sections of the body in Figs. 1, 3 and 4 at any place desired, as, for example, sections 1, 2, 3, 4, 5 and 6.

Take the amount of turn-under at each of these sections from the side turn-under at Fig. 3 and the back turn-under at Fig. 1, and transfer same on to the lines AQ and MO, respectively, Fig. 2. For example, take the turn-under at III, Fig. 3, from the square line R, to convex turn-under line P, and lay it out from point A to III on line AQ, Fig. 2, and so on.

On Fig. 5 is laid out the proportional scale used in finding the sections around the corner between Ag and Mh, Fig. 2. Let X-T, Fig. 5, represent the base line or X-T on Fig. 2. From line X-T, Fig. 5, first lay out the side turn-under of the body, g to A, taking same from g to A, Fig. 2.

At any convenient distance from gA, Fig. 5, lay out the amount of rear turn-under, h to M, from line X-T. Draw a straight line through A-M; likewise lay out sections 1, 2, 3, 4, 5 and 6 on side and back turn-unders, Fig. 5, as taken from Figs. 1 and 3, and connect with straight lines.

Transfer the turn-unders between lines Z-Y and X-T, Fig. 2, at B, C, D, E, F, G, H, I, J, K and L, laying same out on proportional scale from line X-T until they strike line Z-Y, Fig. 5. As the greatest amount of turn-under on this corner is between A and M, Fig. 2, we have to extend lines X-T and Z-Y beyond M on the proportional scale, Fig. 5.

Having the proportional scale complete, it only remains to lay off the different turn-unders, 1, 2, 3, 4, 5 and 6, at B, C, D, E, F, G, H, I, J, K and L from the scale on to the corresponding points of Fig. 2.

POSITION OF PROPORTIONAL CORNER AND LAYING OUT OF HORIZONTAL SECTIONS AND CORNER STRAINERS ON A TORPEDO BODY—Continued

This much accomplished we will connect the points between Z-Y and X-T on lines B, C, D, E, F, G, H, I, J, K and L, Fig. 2, and we will have the desired sections of the corner at the heights of 1, 2, 3, 4, 5 and 6, Figs. 1 and 4.

On most bodies of the type illustrated here, the construction calls for a strainer set up and framed into the sill at the corner, as shown in Fig. 2. This strainer should be set up square from the bottom, that is, square from a straight line passing through the intersections of the outer edges of the strainer and corner at the bottom, in order to obviate as much as possible the necessity of beveling the strainer. In this problem we illustrate a very simple method for laying out the full size of the strainer at the front and back sides.

First let us lay out the strainer on the corner of the body, Fig. 2, wherever it may be desired. Take the width of the strainer and lay it out on line X-T at ab, Fig. 2. From a straight line passing through these points, square front and back lines of the strainer through sections 1, 2, 3, 4, 5 and 6 until they cut line Z-Y at cd, this being the top of the strainer in Fig. 2.

Having determined the front and back lines of the strainer on Fig. 2, proceed to lay same out in its full size as Figs. 6 and 7. Let us lay out a straight line W perpendicular to the front face a-e of the strainer from the point c. This line should be equal in height to line N from T to Y in Fig. 1.

At right angles to this line, draw lines 1, 2, 3, 4, 5 and 6, taken at the heights of these lines on Fig. 1. Wherever the sections 1, 2, 3, 4, 5, 6 in Fig. 2 intersect the front line of strainer, square points on to lines 1, 2, 3, 4, 5 and 6, Fig. 6. Connect these points and it will give us the outside turn-under line V of the strainer. For the inside line U, simply gauge the thickness of strainer from line V or the outside, Fig. 6. S, Fig. 6, shows the line of the stock from which the strainer would be cut.

The same operation is applied to the laying out of the back side of the strainer, Fig. 7, except that all points should be squared from the back side of strainer or line bd, Fig. 2. I-1, IV-4 and c-d are sections of the strainer taken from corresponding lines on Figs. 1, 2 and 4.

Laying Out of Twisted or Winding Surfaces, Illustrated on a Torpedo Body

WE INTRODUCE in this problem the application of a quick and accurate method for laying out twisted or winding surfaces on the side of a torpedo body whereby the proportional horizontal and turn-under sections of the body can be obtained so that after the body is constructed it will possess harmonious and symmetrical lines. It will be noted that in the illustration of this body there is a gradual twist from the rear corner to the dash, Fig. 2. In order to lay out the turn-under sweeps at each of the standing pillars D, C and B, Fig. 1, it is necessary to obtain horizontal sections through the body at 3, 4 and 5, Fig. 1. (First the body should be laid out complete as possible, independently of the horizontal sections 3, 4 and 5. The top line 2 at the waist and the base line 6 should be laid out in Fig. 2.)

Refer to Fig. 3, which shows the side turn-under line of the body at A and the application of the proportional triangle to obtain the shape of the turn-under line E. Having assumed turn-under line A, take the amount of turn-under at E, Fig. 2, and transfer on to line VI from Q on line R, Fig. 3. Lay out section lines III, IV and V, in Fig. 3, at the same heights as corresponding section lines in Fig. 1. Lay out the triangles Q-A-O and Q-E-O from the base line VI, Fig. 3, the apex of the triangles O being located at a convenient distance below VI on line R. Wherever the sections cut the turn-under sweep A on Fig. 3, take section V for example, square the intersection from x on to the large triangle at y. From y draw a perpendicular line until it strikes small triangle at z. Square point from z on to the section line V again, and this gives us point through which the turn-under sweep E passes. Continue this operation with all the sections until the desired points are obtained by which the completed sweep E is determined.

Now having the turn-under sweeps at A and E, let us next proceed to lay out the horizontal sections at 3, 4 and 5 through the body in Fig. 2. In order to establish these sections we will bring into use the proportional scale, Fig. 4. Let e, Fig. 4, represent the turn-under of the body at E, Fig, 2, and a the corresponding turn-under at A in Fig. 2. Lay out on e, Fig. 4, turn-unders at sections 3, 4 and 5, as taken from III-IV-V, Fig. 3. Likewise lay out the turn-unders on a, Fig. 4. Lay out straight lines between a and e at 3, 4 and 5, Fig. 4, and take the total turn-unders at D, C and B, Fig. 2, and lay them out on proportional scale in Fig. 4, at d, c and b. From this we can readily prick off our turn-unders at 3, 4 and 5, from d, c and b, and transfer the same on to D, C and B, Fig. 2. These points established, it remains only to connect them with proper side sweeps and we have completed the horizontal sections, 3, 4 and 5, from which, with lines 2 and 6, we can lay out the exact turn-under sweeps of the standing pillars at D, C and B, Fig. 1.

Fig. 5 is the proportional scale used in laying out the sections around the corner, Fig. 2. Fig. 6 is the proportional scale for laying out the sections through the shroud. Figs. 7 and 8, the one-half rear and one-half front views, respectively, are usually worked out after the turn-unders are determined.

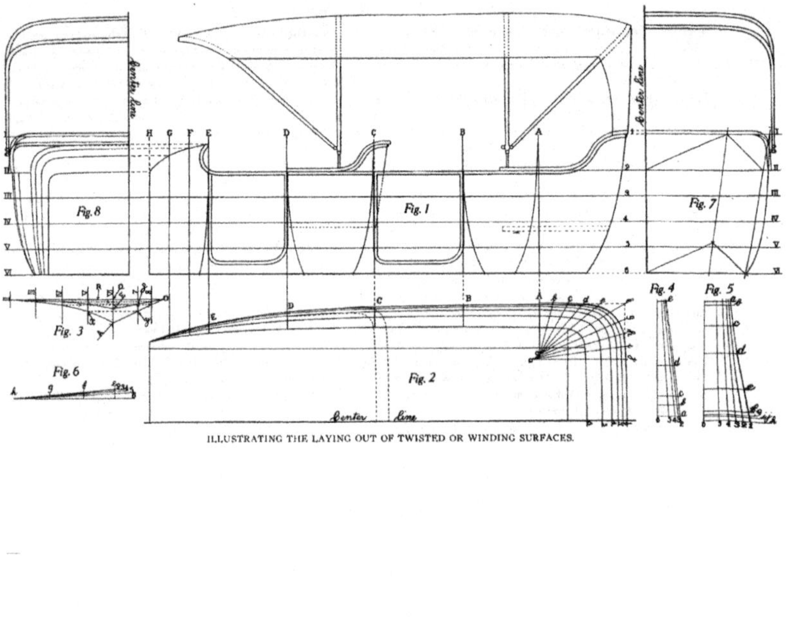

ILLUSTRATING THE LAYING OUT OF TWISTED OR WINDING SURFACES.

Laying Out a Belt Line and Moulding for a Limousine Body

A TRUE and graceful belt line around the corner and back of a limousine body is sometimes difficult to obtain. To produce a working line on the draft which we may feel certain will give satisfactory results is what we will endeavor to explain and illustrate in this problem. It is essential that the belt line on the back of the body should have a generous arch or sweep (usually the same as the roof), and to accommodate this, the belt line at the center of the back should be from 6 to 8 inches higher than the belt line at the lowest point on the side of the body, this, of course, depending somewhat on the width and design of the body, etc.

Determine the position of the belt line on the side of the body and at the center of the back on Fig. 1. Between these two points draw the horizontal section lines II, III, IV, V and VI, Fig. 1. Next lay out these intermediate sections on the corner in Fig. 2. To produce these sections in Fig. 2 we will bring into service a system which has been explained in previous problems, that of the proportional scale, Fig. 3.

With a suitable sweep or curve, draw the line on the semi-rear view, Fig. 4, from 1 to 3, number 3 being the point from which the corner in Fig. 2 commences to round. This back sweep will intersect the horizontal section lines 2 and 3, Fig. 4. Take the distance from the center line of the body to each intersection, and transfer same until they strike the sections 2 and 3 in Fig. 2. Next project points 2 and 3 from Fig. 2 on to Fig. 1 at corresponding section lines. Connecting points 1, 2 and 3 in Fig. 1 will give us the line of the belt corresponding to the line intersecting points 1, 2 and 3, Fig. 4.

There now remains to be laid out the continuation of the belt line from point 3 to a in Fig. 1. This is worked out by the eye, and care should be taken to produce a graceful and true line. After the belt line on the profile in Fig. 1 is completed, project the same on to Fig. 2, and thence to Fig. 4. The line which we have established on Fig. 1 does not look as pleasing to the eye as would be desired, but, on the other hand, after the body is constructed and the line is viewed in perspective, it will be surprising to note what a graceful and well proportioned belt line this method will produce. It must be remembered that the corner of the limousine body is never viewed as it appears on the working drawing, but is always seen with a certain amount of perspective, and appears very much different than it does on the draft.

On metal paneled bodies the panels are joined at the belt line, and are usually covered here by a metal moulding, which is securely fastened to the body framing. This moulding is sometimes of soft metal, and sometimes of steel. If it is of soft metal it is not very difficult to form the moulding around the belt line on the job, but if it is a steel moulding that is used, it is desirable to bend or form the moulding flatwise first, and in order to do this a line should be laid out full length on a board or some indestructible article for the blacksmith, to be used in forming this moulding. To obtain the correctly developed line from the working draft, proceed as follows:

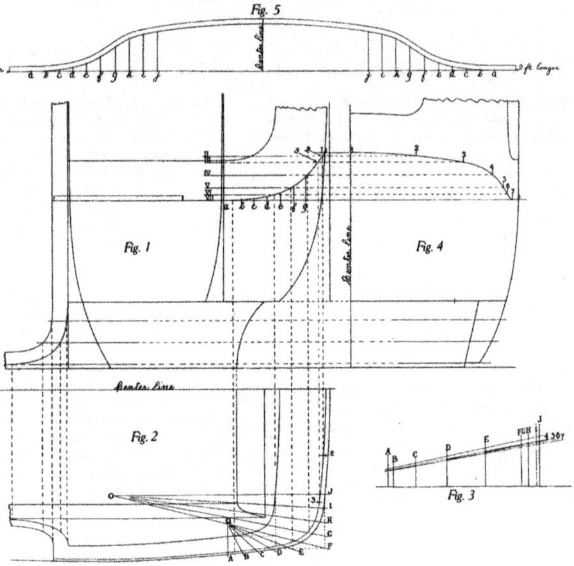

LAYING OUT A BELT LINE AND MOULDING FOR A LIMOUSINE BODY

LAYING OUT A BELT LINE AND MOULDING FOR A LIMOUSINE BODY—Continued

Lay out a long horizontal line, as shown in Fig. 5, and also a perpendicular center line. Square from the horizontal line points a, b, c, d, e, f, g, h, i, j, the distance between each point and from the center line is taken from corresponding points around the belt line in Fig. 2. From the horizontal line in Fig. 1, at the bottom of the belt moulding, prick off the heights of the belt moulding at points b, c, d, e, f, g, etc., and transfer on to corresponding lines in Fig. 5. Connecting these points in Fig. 5 will give us the true shape of the belt moulding on the under side, when laid out flatwise. Gauge the top line of the moulding from this line.

In Fig. 5 it will be seen that the ends of the moulding are broken off short on account of lack of space, the moulding being marked two feet longer at each end. From this point on, the moulding is straight, and it is not necessary to lay this out as long as we take the correct length from Fig. 2.

Pricking Off Corner Pillar on Twisted or Winding Surfaces of a Landaulet Body with Example for Laying Out Joint Irons

LAY OUT the profile of the body as in Fig. 1. Also draw the outside line of the body Y on the bottom view. Then lay out parallel to line Y the bottom line of the body N-O, which for a distance is an imaginary line, but is useful in laying out the cheat line. Determine the amount of cheat wanted and lay out on the bottom view of the drawing as line Q. This is also an imaginary line, considering that the body side continues down to the base line H on the elevation. The adopted cheat line Q creates a twist in the side quarter of the body from vertical section line 8. This is invariably necessary on jobs of this kind for the purpose of producing graceful lines on the back view of the corner in Fig. 4. To determine the amount of cheat wanted, it is sometimes necessary to experiment until we work out what we think is the most pleasing line for the back view of the corner in Fig. 4. After the correct amount of cheat is found, we may adopt same as a standard for all jobs of this style so long as the body side sweeps and widths are uniform; therefore the line is something which cannot be developed, but must be assumed, and depends entirely upon the taste of the designer, and to execute properly requires considerable experience and skill.

To prick off the rear corner of this body, take into consideration the twist or wind in the side surface, and proceed as follows:

In Fig. 2 lay out the normal turn-under sweep of the body at section 8 as shown by the inside turn-under line O. Take the amount of turn-under from the extreme outside of the corner in Fig. 1 to the cheat line Q at the rear, and transfer on to line h-H, Fig. 2. From each turn-under as located on line h-H, Fig. 2, lay out the proportional triangles, the common apex for same being located on line X.

To produce the correct proportional turn-under at the extreme rear corner of the body, proceed as follows:

The normal turn-under sweep O being assumed in Fig. 2, space off on same a number of lines B, C, D, E, F and G at will. These lines also should pass through the rear corner line in Fig. 1. At the point where each line strikes the normal turn-under sweep in Fig. 2, drop vertical projections until the proportional triangle for the normal turn-under sweep is intersected, and from these points draw lines horizontally striking the proportional triangle for the cheat turn-under sweep, and where same is intersected, erect vertical projections on to corresponding lines a-A, b-B, c-C, d-D, e-E, f-F and g-G.

For example, take point on turn-under above line h-H, through which line d-D passes in Fig. 2, and drop a vertical line on to the line of the proportional triangle on line d-D below line h-H. Square this point on to the outside line of the proportional triangle, and from this point erect a vertical projection until line d-D above h-H is again reached. This produces a point through which the proportional turn-under sweep of the body at the rear corner must pass. Likewise proceed with all the other points, and lay out the turn-under sweep Q in Fig. 2. Where lines B, C, D, E, F, G and H, Fig. 1, cut the line of the corner pillar on the profile, project points 2, 3, 4, 5, 6, 7 and 8, passing through the bottom view of the body.

To produce the line of the corner pillar on the bottom and back views, proceed as follows:

PRICKING OFF CORNER PILLAR ON TWISTED OR WINDING SURFACES OF A LANDAULET BODY WITH EXAMPLE FOR LAYING OUT JOINT IRONS—Continued

Take the amount of normal turn-under O in Fig. 2, and lay it out from the horizontal line h-H to a on line 8, in Fig. 3, the proportional scale. At any convenient distance from line 8 lay out from the horizontal base line h-H on line 1 the amount of turn-under with the cheat as taken from line h-H in Fig. 2. This will establish point A in Fig. 3, which should be connected by straight line with a on line 8. Take the amount of over-all turn-under on lines 2, 3, 4, 5, 6 and 7, on the bottom view of body, and wherever these distances fit between lines a-A and h-H of the proportional scale in Fig. 3, erect vertical lines passing through the scale. Continuing, take the amount of turn-unders at lines b-B, c-C, d-D, e-E, f-F and g-G, Fig. 2, from vertical line X to the normal turn-under sweep O, and transfer these turn-unders on to line 8 from a, Fig. 3.

Likewise lay out points B, C, D, E, F and G on line 1, Fig. 3, as taken from vertical line X to cheat turn-under line Q in Fig. 2. Connect corresponding points on lines 1 and 8 in Fig. 3 with straight lines.

On vertical line 2, Fig. 3, take the amount of turn-under from line a-A to line b-B, and transfer from the outside of the body Y on to corresponding line 2 in Fig. 1. Continue the operations with 3, 4, 5, 6 and 7, Fig. 3, being sure to take the correct amount of turn-under each time from the lines which correspond to those in Fig. 1.

Having obtained points on lines 2, 3, 4, 5, 6 and 7 in Fig. 1 through which the bottom line of the corner pillar must pass, proceed to transfer these points on to the rear view of the body, Fig. 4, so as to determine the shape and width of the rear corner pillar in the back view. In order to do this, work from the center line of the bottom view, Fig. 1, to each point on lines 2, 3, 4, 5, 6 and 7, and wherever the corner pillar intersects same lines, transfer distances from the center line in the back view, Fig. 4, to corresponding lines B, C, D, E, F and G. The desired points obtained, connect with graceful and pleasing sweep or curve, and we will have worked out accurately the back view of the corner pillar, from which the stock for same may be laid out with assurance that it will work out correctly.

In laying out the joint irons for the top, the principal object in view is to illustrate an accurate system for determining the center or the break of the joints. After having designed the joint irons when in upright position, and having located the props on the profile of the body, Fig. 1, lay out a horizontal line from the center of the lower prop as shown by dot and dash line on the illustration. Take the distance from point II the center of the lower prop iron to the center of the upper prop I as a radius, and describe an arc striking the horizontal line at III. Determine the position of the center line of the upper prop I after the top is lowered, and with the compasses fixed at point I, or the center of the upper prop, describe an arc V, the radius of which is greater than half the distance from I to III. With the same radius, describe arc IV from III intersecting the arc previously described. Lay out the center line VI, which passes through the horizontal line on which is located points II and III. This intersection determines the center or break of the top joints. From the center of the lowered prop I, to the center of the joint or break on line II-III, describe an arc striking the joint iron when in an upright position, and lay out the center a little bit above a straight line passing through points I and II. The center of the joint or break should always be offset in this way in order to tighten or lock when in an upright position.

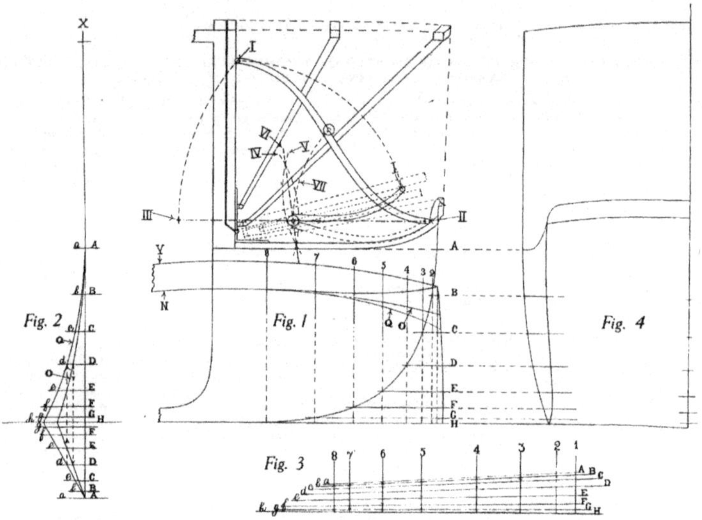

PRICKING OFF CORNER PILLAR ON TWISTED OR WINDING SURFACES OF A LANDAULET BODY WITH EXAMPLE FOR LAYING OUT JOINT IRONS.

The Laying Out of a Coupé Pillar and Construction of Forward Part of an Enclosed Body

IN THIS article our object is to teach the student how to lay out a coupé pillar in detail, which makes a very interesting and instructive problem. In addition we have illustrated and taken up briefly the ordinary construction for the forward part of an enclosed body. This will be especially interesting to the beginner, as it will give him a better insight into the details of construction.

After having decided upon the side view, turn-under and bottom view of the body, we will proceed to lay out the shut bevel of the coupé pillar. The bevel of the pillar where the lock is located should be made to conform as much as possible to the bevel on the face of the lock. From this point to the top of the pillar the bevel is the same.

Lay out section lines II, III, IV, V and VI, Fig. 1, and corresponding section lines 2, 3, 4, 5 and 6, in Fig. 3. Determine upon what radius the door will swing to open at sections 2, 3, 4, 5 and 6. At 5 the bevel is the same as that on the lock, and the door will open without any trouble, but we cannot carry this one bevel on the pillar below section 5, as the turn-under carries the door opening farther away from the center of the hinges O, necessitating a greater bevel in order to have the door open. By trying the shut bevel at each section in Fig. 3 with the trammel points it will be found that the greatest bevel will be at the bottom of the door. Set the trammel points from O to the rear outside edge of the coupé pillar in Fig. 3 at each section, and describe an arc cutting the inside and outside edges of the door, and connect intersections with a straight line, thus establishing the bevel on which the door will open at each section.

In order to obtain a perfect fitting door, the lock pillar and coupé pillar must be worked off to these different bevels, causing the inside face of the coupé pillar to be on a twist below the lock. Notice that the bevels of the coupé pillar are worked each time from the rear outside edge of the coupé pillar. This gives us a straight line for the front of the door and rear of coupé pillar in Fig. 1.

Referring to Fig. 3, A, B, C and D are points on the outside of the coupé pillar at sections 2, 3, 4 and 6. Points a, b, c, d are points on the inside of the coupé pillar at corresponding sections.

From the door line to the inside of the coupé pillar same should be rabbeted out 1/4 inch to allow for the trimming. This is shown on Fig. 1 at sections II, III and IV, also on Fig. 4 at section VI.

In connection with Figs. 2 and 4 we have shown the header, which should be securely framed into the coupé pillar by means of a mortise and tenon joint, as illustrated in section 10. p in Figs. 1 and 4 represents the head rail or roof rail, into which the coupé pillar should be framed by means of a mortise and tenon joint, as illustrated more clearly in Fig. 2 and section 10.

Between the two coupé pillars there are two drop windows which require a center post, Figs. 5 and 6. This center post

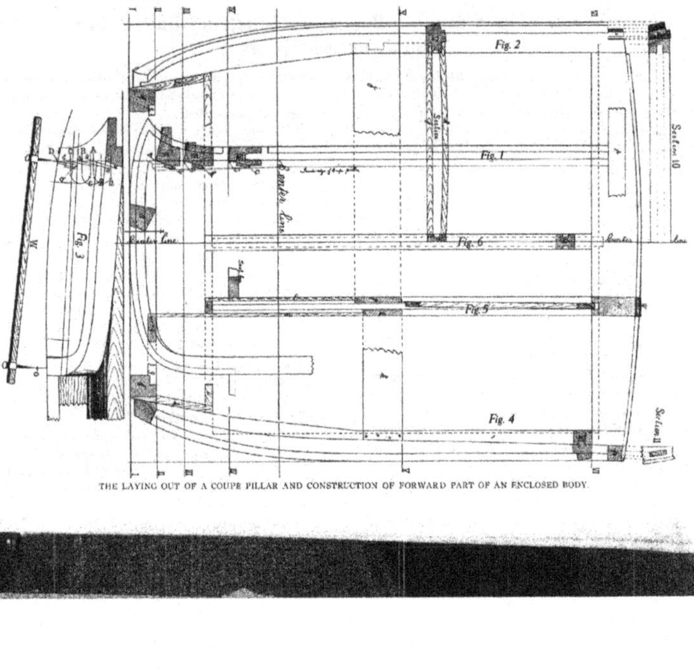

THE LAYING OUT OF A COUPE PILLAR AND CONSTRUCTION OF FORWARD PART OF AN ENCLOSED BODY.

THE LAYING OUT OF A COUPÉ PILLAR AND CONSTRUCTION OF FORWARD PART OF AN ENCLOSED BODY—Continued

is framed into the header by means of a mortise and stump tenon joint, and is lapped into the cross-bar i, Fig. 5.

j, Figs. 2 and 5, represents the neck-bar, which is mortised into the coupé pillar as shown in Fig. 2. Sometimes it is preferable to half-lap this into the coupé pillar.

k, Figs. 4 and 5, represents the garnish board on the inside of the body and on to which is secured the trimming and also the garnish moulding. This should be in two pieces and so arranged as to give clearance enough for, and facilitating the removal of, the window frames. The garnish board should be half-lapped into the coupé pillar, and screwed, but not glued.

The neck-bar j and the garnish board k, Fig. 5, are grooved out to receive the lining boards l and m, which are secured at the bottom to the cross-bars i and g, Fig. 5.

h in Fig. 2 represents a post framed into the sill f and the cross-bar i. This post forms a bearing for the inside of the coupé pillar, which is secured to the same by means of glue and screws.

In Fig. 5 we illustrate a light construction for the division window posts, that is, of fastening ½-inch x ½-inch angle brass to the bottom edge of the glass frame, and permitting one web of this angle to straddle the fence iron instead of having the complete frame to jump the fence, as is customary in most cases. By the use of this construction we lighten the division post in Fig. 5 at least ⅝ inch, or the thickness of the frame.

p, Fig. 4, shows the roof rail rabbeted out to receive the roof board, which is usually of three-ply veneered stock, and should be glued securely and nailed frequently. Over this is glued the roof cloth, which should extend far enough outside of the roof to cover the joint between the roof board and the roof rail.

Section V shows section of the coupé pillar neck-bar and garnish rails at the fence line. It will be noted that the garnish board k in section V is in two sections, and is swept so as to give clearance for window pull when the window is lowered or raised.

The Framing Up of a Door and Pricking Off of a Coupé Pillar

THIS ARTICLE supplements the laying out of a coupé pillar and construction of forward part of an enclosed body. There are a great many different methods for framing up doors. Every bodymaker has his own ideas in regard to this subject. Let this article be intended more for the apprentice than for the experienced bodymaker.

a and b, Fig. 1, represent the door pillars, a being the hinge pillar, and b the lock pillar. The thickness of the hinge pillar is $2\frac{1}{8}$ inches, while that of the lock pillar is about $2\frac{1}{4}$ inches on the outside. These thicknesses are sometimes made more or less, according to the style and construction of the body. e and f are the top rails; g is the lock board; c and h are the fence and belt rails, respectively; j is the lining board; and i is the bottom board. In grooving out the door pillars for the glass frame runs, it is necessary that these be grooved out square. d is the bottom rail of the door, and is half-lapped into the lock and hinge pillars, the joint for the hinge pillar being shown at a and d.

Fig. 7 shows two methods for framing the outside top rail e into the door pillars. Either method will be found practical. The joint used for framing the top rail e into the hinge pillar a is that of a bastard tenon. In framing the top rail e into the lock pillar b we use a stump tenon. The inside rail f is rabbeted into both pillars. Great care should be taken in framing the outside top rail e into the pillars, being sure to get a tight-fitting joint and as much stock outside of the tenon as practical.

Fig. 3 shows a section of the door at the fence line, and the method used for framing a fence rail and lock board into the door pillars. The fence rail c is tenoned into the pillars a and b. The lock board g is halved into the pillars a and b. The belt rail h is also tenoned into the two door pillars. The bottom board i is half-lapped into the door pillars a and b. In this construction it is not necessary to frame the bottom side d into the door pillars, the bottom board i giving all the strength required.

Fig. 6, the section taken at the rear of the lock pillar b, shows the glass frame run and cross-sections of all bars. In laying out glass frame runs, care should be taken to have plenty of clearance for the glass frame to drop easily without sticking and at the same time avoid rattling. The glass frames are $\frac{5}{8}$ inch thick, and the runs are 11/16 inch wide at the fence and top and bottom of run. This allows for clearance, and also paint. It is always well to lay out the width of the door pillars at the fence line first; $\frac{3}{8}$ inch from the outside of the pillar to the window seat is sufficient. Allow 11/16 inch for the window seat, and $\frac{1}{8}$ inch for the fence plate, $\frac{1}{8}$ inch clearance, another 11/16 inch for the glass frame run, $\frac{1}{8}$ inch more clearance, and $\frac{3}{4}$ inch for lock board, making the width of the door pillars at the fence line $2\frac{7}{8}$ inches, in Fig. 6.

Determine the length of the glass frame, which should project above the lower edge of the top rails e and f, Fig. 1, about $\frac{3}{8}$ or 7/16 inch, and lay out the length of the glass frame run below the fence line in Fig. 6 so as to allow the glass frame or window to drop into the door. It is well to

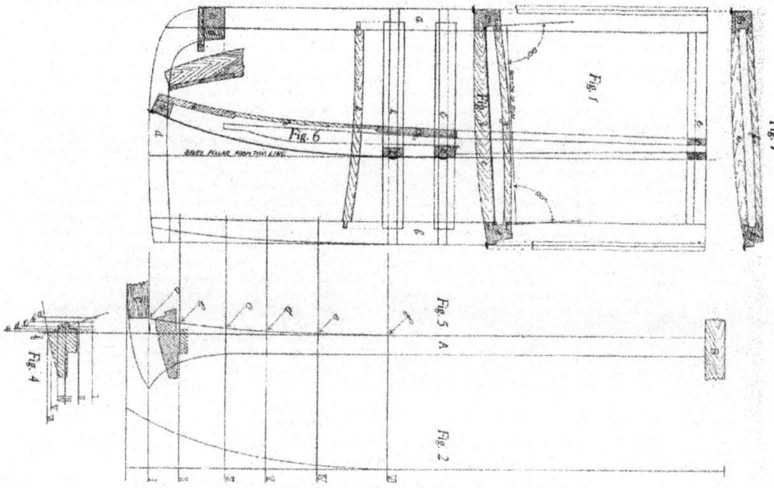

THE FRAMING UP OF A DOOR AND PRICKING OFF OF A COUPÉ PILLAR.

make this calculation so that about $\frac{1}{4}$ or $\frac{3}{8}$ inch of the top frame of the window will show above the fence plate, although in some cases it is impossible on account of the conditions of the chassis, sometimes necessitating cutting the corner of the door, which prevents the window from dropping clear down. At the lower end of the glass frame run in Fig. 6 note that the run is 11/16 inch wide, and parallel for a short distance. This causes the window to fit closely at the bottom, and prevents chattering at this point. The glass frames should never come in contact with the window runs, there usually being screwed channel rubbers on the ends of the frame which work up and down in the runways. There should be plenty of clearance between the door bars, lock board, lining boards and the glass frames. As there is usually attached a strap to the bottom of the glass frame for raising the window, it is always advisable to sweep the lock board g in Fig. 3, which gives clearance for this strap and fixture. Set the lock board and lining boards so that they may be easily removed in case of any interference on the inside of the door. The lock board should not be glued to the pillars, but just screwed, and the joints should be well painted. Always give the runways a coat of linseed oil or primer as soon as possible.

Referring to the framing up of the top and fence rails, we advise that the rails be recessed or set under 1/16 inch, as shown in Fig. 6, thus breaking the joint.

These are a few general instructions and hints which may apply to the framing up and construction of any high door, but should not be regarded as fixed rules.

With reference to the pricking off of a coupé pillar, this is shown in Fig. 5 at A. Fig. 2 is the turn-under of the pillar, and B is the top rail. C is the bottom side. Fig. 4 represents a section of the coupé pillar, and shows one shut bevel throughout.

In getting out a coupé pillar with one shut bevel from top to bottom, that is, beveling the pillar from a plumb line drawn from the outside of the pillar before it begins to turn under, causes the coupé pillar to be heavier at the bottom. This is a simple method for getting the shut bevel of a coupé pillar, and has been employed in carriage work where the turn-under of the body was not very great; but on automobile work, where the bodies are a great deal wider, necessitating more turn-under, it is not found as satisfactory as the method explained in the problem for laying out a coupé pillar and construction of forward part of an enclosed body, inasmuch as it produces a curved line passing through points a, b, c, d, e and f, Fig. 5, which causes the door to be narrower at the bottom than it is at the top.

In pricking this pillar off, lay out the horizontal sections I, II, III, IV, V and VI through Figs. 2 and 5. Take the amount of turn-under at each section from Fig. 2 and lay it out on Fig. 4, passing through the shut bevel of the coupé pillar. Wherever these sections intersect the shut bevel at a, b, c, d and e, Fig. 4, project points until they intersect corresponding sections at a, b, c, d, e and f, Fig. 5. Connecting these points gives us the line for the back of the coupé pillar and door.

The Construction of Glass Frames

A GREAT deal may be said in regard to the construction of glass frames, but in this article we will treat of the ordinary construction, which, like any part of a body, much depends upon the judgment of the mechanic.

Fig. 1 represents an assembled glass frame, a being the bottom piece, b the back, c the top, and d the front. The glass may be removed without injury to the frame.

In order to accomplish this it is necessary to leave two of the joints unglued. In this case we leave the lower joint in the left hand corner and the upper joint in the right hand corner loose so that the frame will be in two pieces. If all the joints were glued up solid, and anything happened to the glass that it should break, it would be almost impossible to reset a new glass into this frame. As it is constructed here, this can be easily accomplished without injuring the frame in any way. The inside edge of the frame is rounded off as shown in Fig. 3, and the frame should be mitered at the inner corners as far as the round, illustrated in Fig. 1.

d1 and d2, Fig. 1, show the lower part of the front piece of frame and also show the shape of the tenon which is let into a mortise in the bottom piece a1. The same kind of a joint is used on all corners except the round corner, and this joint consists of a bare-face mortise and tenon. b1 and b2, Fig. 1, show the lower end of piece b. It also shows the shape of the tenon. a1 and a2 show a part of the bottom piece a which is mortised to take the tenon on the back piece b. Note that the shoulders run horizontally on the lower joints. The frequent raising and lowering of the window causes considerable strain on these lower joints, and in constructing them this way they are much stronger and there is less liability of the joints opening up. If the shoulders were vertical on the lower frame, all the strain would be on the tenons, and the joints would be more liable to open. Notice that the tenon on d1 in Fig. 1 is cut on an angle, which is also true of the tenons on the top piece c.

Another way for constructing the joint at the round corner or joining the two pieces a and b is shown in Fig. 2. In this method we mitre the two pieces together and unite them with a false tenon. This joint is frequently used, but is not as reliable as that shown in Fig. 1 because it shows a longer joint and there is more liability of the joint opening.

Fig. 4 represents a section of a glass frame which is rabbeted out to receive the glass. The rabbet is made deep enough to allow for a moulding which is fastened on the inside of the frame to hold the glass in place. Using this construction it is possible to glue all joints in the frame, as the window glass is fastened and held in place by means of the moulding b.

Referring to Fig. 3, b shows a section of the glass frame grooved out to receive the glass f. In using this construction there should be placed between the glass and the frame a thin strip of rubber, as shown by e. This holds the glass tight in place and prevents rattling, which is sometimes the case when the glass is not of uniform thickness and the glass frames are grooved out uniformly. This rubber is shellaced to the glass. Sometimes nothing but putty is used for setting the glass, but, if rubber cannot be procured, we believe felt will be found to give satisfaction.

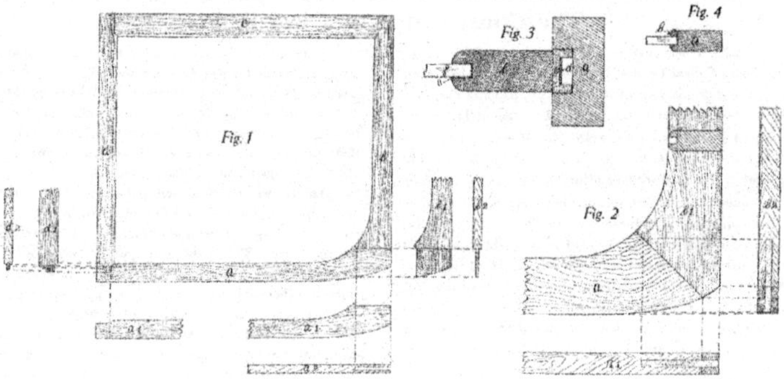

THE CONSTRUCTION OF GLASS FRAMES.

THE CONSTRUCTION OF GLASS FRAMES—Continued

Speaking briefly of the method for setting the glass frame into a job, a, Fig. 3, represents the door pillar grooved out for the frame. On the end of the glass frame is screwed a channel rubber. The plate d is usually set inside the channel rubber c, and the screws are put in from the inside and through the plate d. The channel rubber c should fit closely into the glass frame run in the pillar a, Fig. 3, and the glass frame itself should not come in contact with the glass frame run. Glass frames are usually made of mahogany, and the finish is important; therefore pains should always be taken to obtain perfect fitting joints, and the best of judgment should be used in the construction of the joints, for which there is no law laid down, but the foregoing are simply some practical hints on construction which has proven very satisfactory.

The Laying Out of a Wheel-house and Rear Mud Guard

IN A BODY which is especially wide across the rear seat it is often necessary to make a wheel-house, or depression in the side of the body, so as to give sufficient clearance for the rear wheel. In this event the rear mud guard usually fits into or close to this wheel-house, and conforms somewhat to its shape. We will first take up the method for laying out the wheel-house in a body, such as illustrated.

Lay out the profile of the wheel-house and mud guard on the side view of the body, Fig. 1. On the rear view, Fig. 4, determine the amount of clearance necessary between the body and the wheel, and lay out inside line of wheel-house t-q-x. From Figs. 1 and 4 we will lay out the wheel-house on the bottom view in Fig. 2.

On Fig. 1 lay out on the profile of the wheel-house, points 2, 3, 4, 5, 6 and 7, arbitrarily, and by means of the proportional scale in Fig. 5, lay out sections in Fig. 2 taken at heights of points 2, 3, 4, 5, 6 and 7 in Fig. 1. Squaring these points from Fig. 1 on to sections in Fig. 2 will establish points through which the outside of the wheel-house line r must pass. The inside line of the wheel-house t should be taken from the section of the wheel-ouse at t on Fig. 4, and transferred on to Fig. 2, where it will show as a horizontal line.

Wherever the outer line of the wheel-house intersects the inner line at the front and rear, numbers 1 and 8, Fig. 2, square down points intersecting the profile of the guard, establishing points 1 and 8 in Fig. 1. Lay out horizontal section lines 10, 11 and 12 on Fig. 1. By means of the proportional scale, Fig. 5A, lay out these sections through the inside line of the wheel-house t on Fig. 2, and wherever same crosses the inside line of the wheel-house t drop vertical projections until they strike the corresponding sections on Fig. 1. Connecting these points 10, 11 and 12, with 1 and 8, Fig. 1, will give us the bottom line of the wheel-house on the profile, or the line where the surface of the wheel-house runs out on the surface of the body.

It will be seen on Fig. 2 that there is a ledge r-s about ¾ inch wide around the top of the wheel-house. This should be set into the wheel-house level across the body so as to form a bearing for the mud guard. The mud guard sets under the wheel-house and is secured at this ledge.

In Fig. 6 we have the turn-unders a, b and c, which show the turn-under sweep at corresponding section lines a, b and c, Fig. 2. Fig. 3 is the turn-under at the commencing of the round corner on the side at A, Fig. 2.

After this much is understood it will readily be seen what a simple matter it is to lay out the mud guard. It is first necessary to lay out the profile of the guard in Fig. 1. On Fig. 2 the inside line s of the ledge of the wheel-house will represent the inside line of the guard in Fig. 8 between points 1 and 8. Determine the width of the guard at the rear, and lay out inside line of same in Fig. 8 until it strikes the corner of the body at 9. Connecting 8 and 9 we have the inside line of the guard, Fig. 8, as far as 1. From 1 down, it follows closely the side of the body until it reaches the top of the chassis frame. From this point until it joins the running board it is usually cut away for the spring or brake rods as shown at y-z,

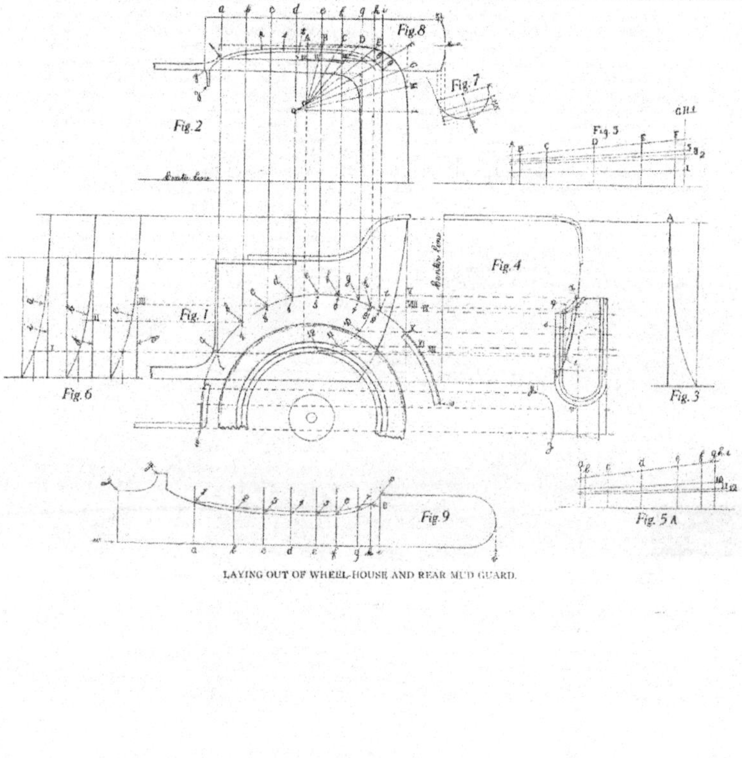

LAYING OUT OF WHEEL-HOUSE AND REAR MUD GUARD.

THE LAYING OUT OF A WHEEL-HOUSE AND REAR MUD GUARD—Continued

Fig. 4. Lines a, b, c, d, e, f, g and h, Fig. 8, are projections from points 1, 2, 3, 4, 5, 6, 7, 8 and 9 in Fig. 1, and will be used in developing the pattern of the guard in Fig. 9.

Fig. 7 shows a section of the rear end of the guard as taken from Figs. 1 and 4 and brought up close to Fig. 8, with the end squared from the profile in Fig. 7, for the purpose of working out the points K, L, etc., on the plan view of the end of the guard in Fig. 8. Let lines K, L, M and N be drawn through the end of the guard until they strike the profile in Fig. 7. Draw vertical lines from each point through the end of the guard on Fig. 8. From the center line on Fig. 7 take the widths of the guard at K, L, M and N, and transfer on to corresponding lines from the center line in Fig. 8. This gives us a true view of the rear end of the guard in Fig. 8, although the same is not necessary for laying out the pattern in Fig. 9.

The sole object for laying out the guard in Fig. 8 is for pricking off the widths at various points to be used in developing the pattern, Fig. 9. To obtain the pattern, Fig. 9, it is first necessary to lay out the full length of the mud guard around the curve of the profile w to v, Fig. 1. Lay out developed length on the outside edge of the guard w to v, Fig. 9. Transfer distance between points w, a, b, c, d, e, f, g, h, i from Fig. 1 to the pattern, Fig. 9, working from the bottom of the guard w to get the true length at each point in Fig. 1. Through these points in Fig. 9 draw lines perpendicular with outside of guard. Transfer widths of the guard at lines a, b, c, d, e, f, g and h, Fig. 2, to corresponding lines from outside of guard in Fig. 9, establishing points 1, 2, 3, 4, 5, 6, 7, 8 and 9. Connecting these points will give us the inside line of the guard where it fits into the wheel-house.

Lay out the round end at the rear of the guard, using the same round or radius as in Fig. 7. Prick off the front of the guard from Fig. 4 on to the pattern on Fig. 9. Add the necessary flanges all around the guard for wiring or whatever may be desired, and the pattern is complete, ready for cutting out the stock.

Laying Out Pattern of Panel for a Metal Seat by the System of Triangulation

TRIANGULATION is a system for measuring up and developing surfaces for the purpose of obtaining patterns for sheet metal work, and has long been used for this purpose by sheet metal workers and cornice makers. It can be applied to the automobile and carriage trade in making patterns of panels with straight flares, but cannot be used on panels that are convex or concave both ways, or, in other words, on panels that have to be hammered. For a convex or "King of Belgium" seat this system would be of no use, but for a straight line seat, as illustrated herein, or for mud guards, shrouds, underbody panels, etc., it is invaluable, inasmuch as it saves the time and expense of setting up temporary forms for obtaining the desired patterns.

Referring to the illustration, Fig. 1 is the elevation of a metal automobile seat, and Fig. 2 is one-half the bottom view. It is from these two views that we obtain the pattern of the panel, Fig. 5.

Before laying out the pattern of the seat it is necessary to establish the line of the arm rail in Fig. 2 as it drops down on the flare of the seat. Space off on the top line around the seat numbers 1, 3, 5, 7, 9, 11 and 13, Fig. 2, commencing where the corner starts to round at the back, and continuing around the seat as shown. Space off as many parts equally on the seat line, Fig. 2, connecting these points on top and bottom lines of seat with straight lines, project same points and lines on to corresponding positions in Fig. 1. Wherever the arm rail line on the profile intersects the oblique lines connecting top and bottom points of seat, Fig. 1, project these points on to the corresponding lines in Fig. 2.

For example, take point VII on line 7-8, Fig. 1, and project horizontally until same strikes line 7-8 at VII in Fig. 2, thus establishing a point through which the arm rail in the bottom view must pass.

Lay out the triangles from these points on the arm rail and points on the seat bottom on Figs. 1 and 2 as shown.

In starting the pattern, Fig. 5, lay out the center line OQ, which is equal to the full height on the flare of the seat at the back, Fig. 1. With the dividers take the distance from Q to 1 on the top line of the seat, Fig. 2, and transfer on to the pattern, using this distance as a radius, and swing it out both sides of the center line from Q.

In order to find the intersection of point 1 on these arcs, proceed as follows:

Use the distance from O to 1 in Fig. 2 as the base of a triangle, and lay same off on any convenient space on the drawing, as Fig. 4, O to C. Taking the height of the seat perpendicularly from the bottom to point 1 at the top as the altitude or perpendicular, lay out on the triangle chart in Fig. 4, C to 1. Connecting O with 1 in Fig. 4 will give us the hypothenuse of the triangle or the true length of line

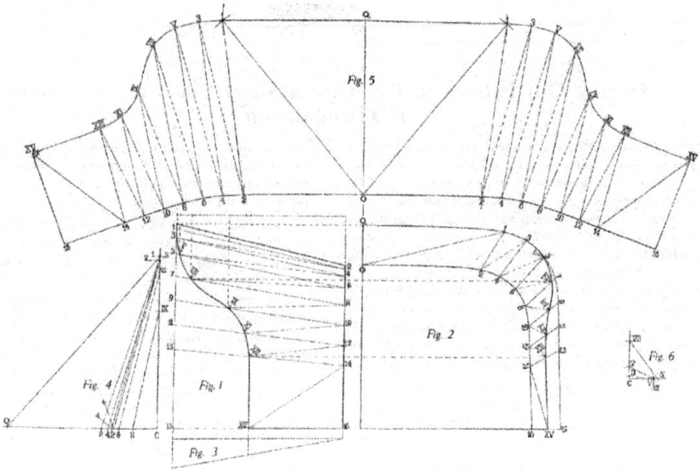

LAYING OUT PATTERN OF PANEL FOR A METAL SEAT BY THE SYSTEM OF TRIANGULATION.

LAYING OUT PATTERN OF PANEL FOR A METAL SEAT BY THE SYSTEM OF TRIANGULATION—Continued

O1, Fig. 2, which distance should be laid out from O on Fig. 5 until it intersects the arc drawn from Q at 1.

To obtain point 2 on pattern, take distance from O to 2, Fig. 2, and lay it off as a radius from O, Fig. 5, both sides of the center. Take the distance 1 to 2, Fig. 2, as a base of a triangle, and lay it off on Fig. 4 from C. Connect 1 and 2 for the hypothenuse or the true length of the line sought, and lay distance out from point 1, Fig. 5, until it intersects the arc drawn from 0 at 2, and so on until the arm rail commences to drop at 3. After 3 on Fig. 2 we cannot take the spaces from the arm rail line direct, as same is dropping until it runs into the horizontal line at XIII.

In order to space the top line of the arm rail correctly from point to point on Fig. 5 it is necessary to lay off another set of triangles to obtain the true length of lines between 3 and XIII, Fig. 2, which is done as follows:

To obtain the line 3 to V, Fig. 5, take the space from 3 to V, Fig. 2, and lay it off from C on Fig. 6 as the base of a triangle. Take the amount that the arm rail drops, from 3 to V on Fig. 1, and lay it off from C on vertical line as the altitude, Fig. 6. Connecting 3 and V will give us the hypothenuse of the triangle c3V, Fig. 6, which is the true length of line 3 to V, Fig. 2, and should be laid off from 3 on Fig. 5. The same operation is necessary for every space on the arm rail between V and XIII, laying out each triangle to obtain the true length of each line.

As the points on the arm rail after 3 are getting lower it is necessary to change the altitude of each triangle on Fig. 4. Take the distance from 8 to IX, Fig. 2, for example.

Using line 8 to IX as a base, lay this distance out on Fig. 4 from C to 8. Take the perpendicular height of IX on Fig. 1 from the seat bottom and lay it out as the altitude, Fig. 4. Connecting 8 and IX, Fig. 4, will give us the hypothenuse of the triangle or the true length of line 8 to IX, Fig. 2, which should be laid out from 8 on Fig. 5 until it intersects the arc drawn from VII at IX. Continuing these operations with all the points as shown will give us the outline of the pattern, Fig. 5. Around this pattern whatever flanges are necessary for turning under the seat frame and railing, etc., may be added.

Laying Out of Pattern for a Shroud Panel

THE SAME system of triangulation is applied in this problem for obtaining the pattern of a shroud as was used for working out pattern of the metal seat panel.

The elevation and front view of the shroud, Figs. 1 and 2, are first laid out. Then on Fig. 2, the front view of the shroud, space off points on inner and outer lines as indicated by numbers. The location of these points is not fixed by any rule, but good judgment should be used in placing them where necessary only, the corners being the parts where the spaces should be laid out most frequently in order to insure the most accurate results. The flat or nearly flat surfaces do not require as many triangles. Connecting the points on the inside and outside lines of the shroud in Fig. 2 will establish the triangles with which we will lay off the full pattern of the shroud in Fig. 5.

Lay out the full length of line O-Q on the center of the pattern, Fig. 5, as taken from Fig. 1. In spacing off the points on the back line of the pattern in Fig. 5, do not take the distances direct from the outer line on Fig. 2, as these spaces do not represent the exact length of the lines, as well be seen by looking at Fig. 1. The rear line of the shroud slants back toward the bottom. Therefore lay off triangles for each space separate, as in Fig. 6.

For example: In spacing off point 1 in pattern, Fig. 5, take the distance from 1 to Q in Fig 2, and lay it off as the base of a triangle on line a-a, Fig. 6, from Q. Take the distance that the shroud slants back, Q to 1 in Fig. 4, and lay it out in Fig. 6 from line a-a to point 1 as the altitude of the triangle. From Q to 1 in Fig. 6 will be the hypothenuse or the true length of the line Q to 1 in Fig. 2, which should be swung off both sides of Q in Fig. 5.

In order to find point 1 in Fig. 5, obtain the true length of line O to 1 in Fig. 2, as follows:

Square upon line a-b in Fig. 4 point 1 from Fig. 1. Take the distance from O to 1 in Fig. 2, and lay it off from line a-b at the front line of the shroud in Fig. 4. Connect O and 1, and we will have the hypothenuse of the triangle a-O-1 in Fig. 4, or the true length of the line O to 1 in Fig. 2. Lay this distance off from O in Fig. 5 until it intersects the arc drawn from Q at 1.

To describe the arc from O, Fig. 5, on which point 2 will be located, take distance directly from O to 2 in Fig. 2, using same as the radius for describing arc in Fig. 5. As the front line of the shroud is perpendicular with the base line, all even-numbered points can be taken directly from the front view in Fig. 2. For finding point 2 on the pattern, take distance from 1 to 2 in Fig. 2 as the base of a triangle, and lay it off on vertical line from a in Fig. 4. Square point 1 from Fig. 1 on to the line a-b in Fig. 4. From 1 to 2 will be the hypothenuse of the triangle a-1-2, or the true length of the line 1 to 2 in Fig. 2. Lay out this distance from point 1 in Fig. 5 until it intersects arc drawn from O at 2. And so on, it is simply a matter of getting the true length of each line or the hypothenuse of each triangle represented in Figs. 1 and 2, and transfer same on to pattern in Fig. 5.

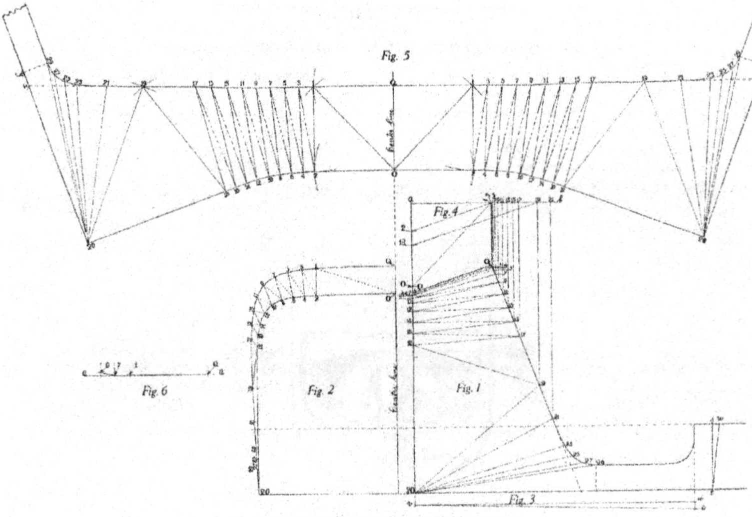

LAYING OUT OF PATTERN FOR A SHROUD PANEL.

LAYING OUT OF PATTERN FOR A SHROUD PANEL—Continued

To avoid complication and confusion we have shown a few of the triangles only in Figs. 4 and 6, but in general practice each triangle will have to be laid out to determine the true length of lines desired. After the descriptions of the operations used in Figs. 4 and 6, with the same principle the remainder of the triangles may be laid out to determine the required points on the pattern, Fig. 5.

C, Fig. 3, represents the contraction line of the shroud at the height of the seat bottom. It has no connection, however, with the pattern, but is useful in establishing the outer line around the shroud in Fig. 2. After the pattern, Fig. 5, is obtained, add whatever flanges are necessary according to the style of construction used, etc.

Laying Out Patterns for a Front Mud Guard and Splasher

IN THIS problem we illustrate and explain methods used in laying out patterns of a front mud guard and splasher, the surfaces of which are twisted. Fig. 1 is the side view or elevation showing front wheel and chassis frame in connection with the fender. Fig. 2 shows the relation of the running board to the chassis and shape of splasher where it meets the running board. Fig. 7 is the plan view of the fender with a full pattern extended.

It is necessary to lay out the side view and plan, Figs. 1 and 7, respectively, and from these two views, with the aid of the triangle charts, Figs. 2, 3, 4 and 6, we obtain the patterns desired. First let us explain method of laying out the full pattern for the twisted mud guard splasher, Fig. 5.

It should be decided which is the most simple way to form the splasher with as little twist as possible, and with this object in view, consider that the surface of the splasher inside of the triangle O-5-Q, Fig. 1, is flat and has but one bevel or flare. Also let us decide that between points Q and A, and 9 and 10, Fig. 1, we shall make the splasher flat.

From point O to points 1 and 5 on the splasher in Fig. 1 it will be necessary to give the splasher a slight twist in order to meet the parallel line of the inside edge of the guard on Fig. 7. Between points Q, 5 and 8 in Fig. 1 it will also be necessary to roll or form the splasher somewhat to meet the inside parallel line of the fender, Fig. 7. The lower end of the splasher where it joins the running board and chassis frame will have to be twisted into the required shape. Wherever swept or twisted surfaces appear it is necessary to space off points and triangles frequently in order to assure accurate and satisfactory results.

From the front of the splasher to the center or the highest point of the splasher, Fig. 1, let us lay out points 1, 2, 3, 4 and 5. Connect these points with the front of the splasher at O on the bottom, and we will have the necessary triangles for laying out this section of the splasher. From point 5 on the top of the splasher, Fig. 1, space off points 6, 7 and 8 until the fender curve runs into the chassis line, Fig. 1. In Fig. 2 we show the shape of the splasher where it connects with the running board. It will be seen that the end of the splasher here is curved. Lay off a number of equal spaces XI, XII, XIII, XIV and XV on the splasher in Fig. 2, and project same points on to the rear end of the splasher in Fig. 1. Connecting these points 11, 12, 13, 14 and 15 with A, Fig. 1, will give us the necessary triangles for developing the surface of this section of the splasher.

On Fig. 2 we indicate the distance from the inside of the mud guard or running board to the outside of the chassis frame, which distance will be the base of all triangles in Figs. 3, 4 and 6.

In commencing the actual pattern, Fig. 5, take the distance from O to Q, Fig. 1, and lay out on the bottom of the splasher in Fig. 5. This is the true length of the splasher between these points. Take the distance from O to 5 in Fig. 1 and lay it out as the altitude of a triangle in Fig. 3. With the distance from the inside of the guard to the outside of the chassis frame as the base of the triangle, determine the hypoth-

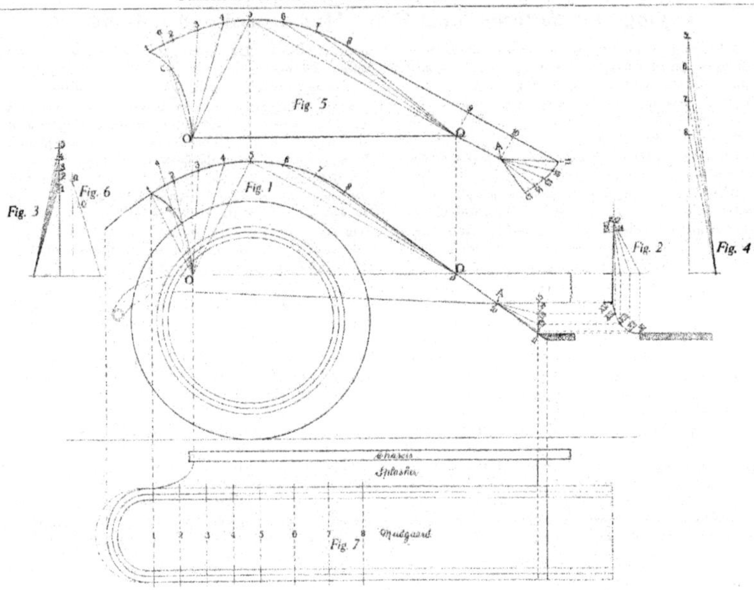

LAYING OUT PATTERNS FOR A FRONT MUD GUARD AND SPLASHER.

LAYING OUT PATTERNS FOR A FRONT MUD GUARD AND SPLASHER—Continued

enuse or the true length of the line O to 5, Fig. 1. Take this distance as a radius, and describe an arc from O, Fig. 5.

To find the intersection of point 5 on this arc, take the distance from Q to 5 in Fig. 1 and lay it out on Fig. 4 as an altitude of a triangle, and with a base equal to the distance from the chassis frame to the inside of the guard, as before stated, find the hypothenuse or the true length of the line desired, and, using this distance as a radius centered at Q, Fig. 5, describe an arc cutting the arc described from O at point 5 in Fig. 5. From point 5 describe an arc with the radius equal to the space from 4 to 5 in Fig. 1.

To find the intersection of point 4 on this arc in Fig. 5, take the distance from O to 4, Fig. 1, and use it as the altitude of a triangle in Fig. 3, and with a base as given, find the hypothenuse and lay it out from point O in Fig. 5 until it intersects the arc drawn from 5 at 4. Likewise space off points 3, 2 and 1, using triangle chart, Fig. 3. Also lay out spaces 6, 7 and 8, using the triangle chart in Fig. 4, and find the intersections of each arc as previously described.

To find point 9 in Fig. 5, take the distance from 8 to 9 in Fig. 1, and lay it out from 8, Fig. 5. The intersections of this arc with a radius drawn from Q, equal to the distance from the outside of the chassis frame to the inside of the guard, will determine point 9 in Fig. 5. From Q to A, and from 9 to 10, simply take the distance direct from Fig. 1, as this is actual length.

For the lower end of the splasher we have spaced off on Figs. 1 and 2, points 11, 12, 13, 14 and 15. Take the distance from 10 to 11, Fig. 1, and strike it out from 10 in Fig. 5. Using the distance from A to 11, Fig. 1, as the altitude of a triangle, lay it off on the triangle chart in Fig. 2 with the base equal to the distance from the chassis to the inside of the guard or running board.

After having determined the hypothenuse of this triangle, or the true length of the line from A to 11 in Fig. 1, lay it out from A, Fig. 5, until it intersects the arc drawn from 10 at 11. Space off the distance from 11 to 12, Fig. 5, taken from XI to XII in Fig. 2.

We now want the true length of the line A to 12 in Fig. 1 in order to enable us to locate the intersection of point 12 in Fig. 5. With the distance from A to 12, Fig. 1, as the altitude of the triangle, lay same off in Fig. 2 with the base obtained by squaring up point XII in Fig. 2 until it strikes the horizontal line at the top of the chassis frame. With the base and altitude given, it is easy to determine the hypothenuse of this triangle, which should be described as a radius from A in Fig. 5 until it intersects the arc drawn from 11 at 12, and so on with points 13, 14 and 15, being careful to obtain the correct base of each triangle in Fig. 2, inasmuch as the base of each triangle is ever changing on account of the splasher running closer to the chassis at each point.

Fig. 7 shows the actual plan view of the mud guard, splasher and chassis, as projected from the side view in Fig. 1. The extended dot and dash lines show the developed surface or pattern of the mud guard as taken by measuring the distance from point to point around the profile of the guard in Fig. 1.

Perspective Drawing of Vehicles

IN THIS article we illustrate and explain a system of perspective drawing for automobiles and carriages. In making perspective designs of vehicles it is always desirable to lay them out so that the dimensions of the body can be scaled from the profile of the drawing.

For instance, if one desires to know the width of the door, the depth of the seat, or the height of the job, he can readily obtain the measurements from the profile of the drawing.

In this case the vehicle will not be set up in true perspective, but the effect is as pleasing as if it were. If the job were drawn in true perspective it would not be possible to scale any measurements from the drawing, and it is not necessary for draftsman or designer to be familiar with true perspective drawing in order to become efficient in carriage perspective. The system explained herein will be found to be very simple, and once the designer masters the system, he will find that it will be of great value to him, inasmuch as he will be able to produce rapid and more pleasing results.

In preparing a perspective drawing, the first operation is to lay out the true side view or elevation of the vehicle, as shown in Fig. 1. Next determine the vanishing point, which in this case is to the front of the coupé pillar nineteen feet on the line of vision, but this distance is taken arbitrarily, and after once determined satisfactorily, may be kept as a standard. The line of vision is usually five feet three inuhes from the base or the ground. The vanishing point is not shown on the drawing, as it would extend far out from the edge of the paper or the ordinary drawing board. Instead of showing the vanishing point, we illustrate a more satisfactory method for laying out the vanishing lines. Fig. 4 represents an arc with the vanishing point as a center. This arc should be made out of celluloid or light wood, and then it may be used at any time by securing same to the drafting board with two thumb tacks, as shown in Fig. 4 by a and b.

Fig. 7 represents a lightly constructed T square with an eccentric blade, that is, the top edge of the blade should be perpendicular with the center of the head. Setting the T square on the arc as shown, we can draw any number of lines that will vanish at the vanishing point on the line of vision.

If the widths of the body are known, it is not necessary to draw the front and back views, Figs. 2 and 3, respectively. This is shown here by way of illustration only, and when the draftsman becomes accustomed to the system, this work may be eliminated to a certain extent.

With T square 7 set on the celluloid arc 4, draw the inclined line A passing through the center of the front near wheel at the base. Let this line continue the full length of the drawing. After this, draw another inclined line B that passes through the center of the near rear wheel at the base. Then drop a perpendicular line from the center of the rear wheel until it strikes the inclined line A at 2, Fig. 8. Considering that the wheel tread is 56 inches, lay out this distance as a radius from 2, describing an arc until it cuts the inclined line B at II, Fig. 8. By connecting these two points with the straight line c we will have the line on which all our cross-widths should be laid out.

PERSPECTIVE DRAWING OF VEHICLES—Continued

Fig. 5 represents a triangle, the angle of which is about sixty degrees, and which angle is used for all penchant lines, and determines the offside points or the amount of perspective. This angle is taken at will, and, of course, regulates the amount of perspective. After the draftsman has made a few drawings he can determine which angle is the most satisfactory and produces the most pleasing drawing, and adopt it as a standard penchant line angle.

To find the off rear wheel, draw the penchant line i with the triangle, Fig. 5, from point 2 on line A, Fig. 8, until it strikes the line B. This gives us the base of the off rear wheel. Lay out the vertical center line, and use the T square Fig. 7 on the arc Fig. 4 for finding the horizontal center and top of wheel.

To lay out the perspective of any part of the vehicle which is wider or narrower than the wheel tread it is necessary to work from the center line, Fig. 8. The center line is obtained by bisecting the wheel tread on line c at x, from which point the center line is drawn with the T square Fig. 7 on arc Fig. 4.

For example, if the perspective on the coupé pillar is wanted, take the width from the center line, Fig. 2. Lay it out either side of the center line in Fig. 8 on line c at V and 5. With the T square Fig. 7 on arc Fig. 4, draw lines from these points V and 5 through Fig. 8. Square down near side of coupé pillar at the front until it strikes line 5. Use the triangle, Fig. 5, to draw the penchant line D until it strikes line V. Erect a vertical line from this point on to Fig. 1, and this gives us the offside of the coupé pillar at the front. Take the T square Fig. 7 on the arc Fig. 4 for laying out vanishing lines and obtaining points on the offside of the coupé pillar which should correspond to points on the near side of the coupé pillar, and so on with 3, 4, 6, 7, 8 and 9, Fig. 8, which represent the widths of the body taken at different points on Figs. 2 and 3, and D, E, F, G, H and i represent penchant lines for obtaining the perspective at these points.

Fig. 9 shows how the vehicle would appear in looking down upon it from above. The near side produces a perfect side view as in Fig. 1. Fig. 9 is, of course, false, as the car could not be twisted as shown in Fig. 9, but this is the way the perspective is set up in the profile or elevation, Fig. 1.

Fig. 10 illustrates approximately how the plan would necessarily have to appear if the elevation or side view, Fig. 1, was in true perspective, and is laid out here by way of illustration to show the difference between true perspective and the system of perspective taken up in this article.

Referring to these two figures, 9 and 10, compare the wheel base on the near side of the vehicle, and it will be noted that the wheel base in Fig. 10 is greater than that in Fig. 9, showing that anything in true perspective set at such an angle will appear shorter than it really is, and, if drawn that way, will not permit of the taking of any measurements from the side of the vehicle, as explained previously. In Fig. 9 the reader will note that the wheel base on the near side is the same as that of the wheel base in the elevation of the vehicle, Fig. 1, therefore making it possible to scale this dimension and others direct from the side view or elevation, Fig. 1.

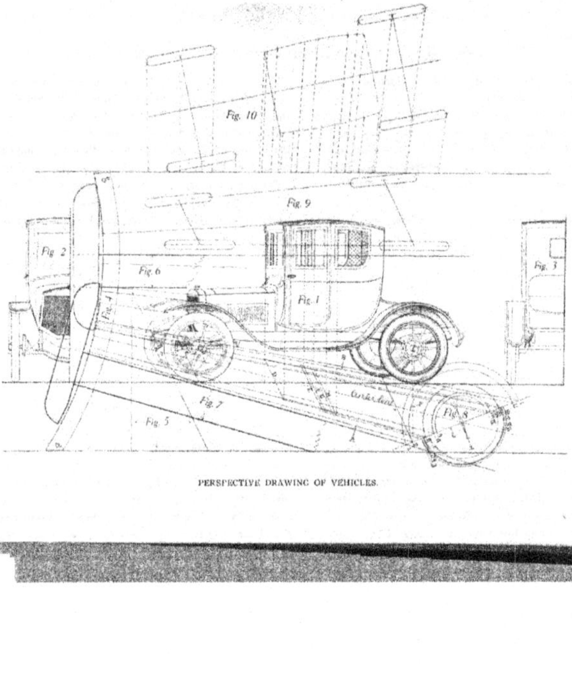

PERSPECTIVE DRAWING OF VEHICLES.

Coloring Carriage and Automobile Drawings

IN COLORING vehicle designs, be they carriages, automobiles, wagons, or, in fact, anything in vehicle construction, a certain knowledge, or, we may say, rules, must be followed to attain the required proficiency to be successful. Besides, considerable practice, after the rules are known, must be had to become a skilful artist in this particular line. Of course, different roads lead to the same end. One road may be longer than the other. Nevertheless, after the rules are known, the artist may pursue his own plan as how to get quickest to the road's end.

Many carriage designers who have taken lessons in color design from experienced artists may differ from our practice and opinions, but, after all, when the first rules have been studied and practiced, success will depend more on practice to obtain perfect work, which is admired even by those who have handled color work for years.

A colored side elevation of any vehicle, carriage, automobile, wagons, etc., drawn in perspective is acknowledged to be the very best representation to show buyers, or to print in catalogues. Drawn in perspective, most of the details contained in the vehicle can be shown, but keeping in mind always that a correct side elevation must be retained to represent the style to advantage. To produce good colored drawings which are admired by others is a great inducement to add this accomplishment to others, which will aid one not only to add taste to his artistic talent, but also furnish an opening to secure a better paying position.

Drawing instruments, the best kind, are necessary, also to have a good variety of brushes, and an assortment of bristol board and other kinds of fine drawing papers, including tracing papers.

In regard to water colors, Winsor and Newton have had a half-century reputation, but a great variety of cake and tube colors can be obtained from stores where artists' materials are sold. The most used colors are white, brown, blue, green, yellow, vermilion, black and deep gray, but, of course, by buying a whole set, other colors are in the box that may be needed. India ink in sticks or bottles is a necessity. Some prefer Higgins' American make, while others prefer the French make, but both are of excellent quality.

Pen drawings, uncolored or colored, should not be drawn directly on bristol board or any other kind of drawing paper, but should be made either on thin white paper or the regular tracing paper. Mistakes in outlining carriages and automobiles are unavoidable, and, therefore, must be erased, and the surface of the very best drawing paper that is made will be damaged with the eraser, and defects show through the colors. After the sketch is done, it is blackened on the rear side, that is, the reverse side from the drawing. For this, dry lamp black is used. Others save the dust from the pencil sharpener. To obtain the dust, they use a small box with a flat file on top. The file is used as a pencil sharpener, and the dust drops in the box, and is kept in the box until needed.

Most artists use red chalk known as "rouge." Some of the black dust will adhere to the paper where not wanted, and it is difficult to remove it, while the rouge is more easily erased. When the sketch is finished and reddened on the rear side of tracing paper, it is put on the bristol board, fastened with thumb tacks, and all the lines are then gone over with a 6-H Faber or Hardtmuth pencil with a sharp point. Others use a needle or tracer, which will be found among the drawing instruments. By this process, a clean, correct reproduction of the design is obtained, and at the same time the draftsman has a copy for future use.

The foreground of the object must be drawn first, which, on a carriage or automobile, means the wheels on the near side. Each object or part of the vehicle is thus drawn, one after the other, and the last part to be drawn will be the off wheels. At the same time, the required colors must always be kept in mind. If the gears are yellow, and striped black, the face of the spokes near the hubs have generally two fine black lines, also one or two fine black lines on the rim faces, and tires black.

In such a case, the hubs, spokes and rims are colored first, and black lines drawn on top of yellow, and all colors are worked same as India ink. For making small lines on gear or body, also

COLORING CARRIAGE AND AUTOMOBILE DRAWINGS—Continued

scrolling, use a lady's pen. In fact, a lady's pen is very handy for all kinds of lines, even the straight lines. Lines of various thicknesses can be made by more or less pressure, thus avoiding the unscrewing of drawing pens.

Working with a lady's pen requires considerable practice, but, when once mastered, it is a great advantage, and considerable time is saved thereby. A colored plate can not be made except all the outlines are drawn with the required color of the ink. If the gear is yellow, the outlines must be of the same color, and if the body is blue, the outlines are blue, if the colors are of a darker shade. Otherwise, if a light shade, the outlines require a fine black India ink line.

To start the color on bodies, if there is black, such as rockers and boot panels, use Winsor and Newton's lamp black, to which, after being rubbed up to a rather thick liquid, add a very little dissolved gum arabic. To know whether the color has the proper consistency when rubbed up, tip the cup. If right for use, the color must flow slowly. To make all outlines sharp, run them over twice with lamp black with the drawing pen, and fill the rest of the panel between the lines. For the small and narrow black spaces, thin down the color somewhat, otherwise it will have a too heavy appearance. For the rest of the panels, that is, the large spaces, the brush is used, but care must be taken to distribute the color properly.

The painting of the body in coach body colors is exceedingly difficult, and requires considerable practice, as very few of the water colors have sufficient body to cover well. Lamp black. Prussian blue, vermilion and Indian are the only ones having good coloring properties. Most artists require transparent colors, while for carriage, coach and automobile work they must cover the surface at the first stroke, as only a few colors will allow a second coat. Carmine and ultramarine are the only ones, and it requires great care and considerable practice to produce a uniform surface after several coats. Perfect work cannot be produced with imperfect colors, but colors at the present time, made just for this work, which cover the surface with but one application, can be bought in first-class stores.

Colors of this kind must be rubbed up to the same consistency as explained for lamp black, and should then be laid on freely, but with extreme uniformity. A spot once dry should not be retouched, as it will show defects on the colored surface.

The striping, which gives finish and relief to a colored drawing, is used either to cover the edges, or on top of the main colors, and is done with the drawing pens, and the small lines with the lady's pen, and one that has been used on other work should be used, as a new pen is too sharply pointed and liable to scratch the colors. The colors for striping are generally ultramarine, vermilion, white, yellow, orange, chrome yellow, cream, canary, or a combination of the above colors.

After the striping lines have been placed on the gears and body, prepare lamp black, which flows well in the pen, and draw a very fine line under, or right and left, as the case may be, representing the shade for the striping line. Shade cushions when they are tufted; lower part darker and upper part lighter. This part can be imitated from clear, well-made photographs and half-tone illustrations. Make handles, lamps, singletree tips white or gold, but as these are generally too thin, they must be prepared to be suitable for this work. Shake the liquid well, pour it into a saucer, and allow it to stand ten minutes; throw off the water which rises to the surface, and mix the remainder with arabic gum water, stir well, and apply to the drawing. When dry, burnish with a thin bone with rounded edges, and shade with burnt sienna.

A great deal more could be said about this subject, but the above explanation is about the whole substance of the matter. In order to be successful, do not mind the tedious labor. Use the greatest care and the very best materials, which are the most essential points.

Working Drafts

THE FOLLOWING working drafts represent nearly every style in automobile bodies, and should be of great service as a reference for designs, etc. They are accurately drawn to a working scale. The different views are kept clear of one another, which makes them easy to follow.

Unfortunately, on account of having to make shop working drafts full size, it is necessary sometimes to lay out one view on top of another.

Good carriage draftsmen and men who can lay out and construct work from the draft are in good demand, and are usually the best paid men in the trade.

To be able to make a complete working draft of a carriage or automobile body requires considerable skill and experience. Our advice to beginners is to practice free-hand drawing and train the eye to obtain good, true lines, not depending on mechanical methods. Scale drawing is very good practice.

The experience obtained by practical work in the body shop is invaluable. This will give the student a better understanding of the information required on a draft.

The problems taken up in this book should be thoroughly understood, so as to be able to apply them correctly. Aim to make the draft as plain and simple as possible, putting on only the necessary working lines, as any unnecessary lines confuse the workman as well as waste the draftsman's time.

The projections and connecting lines shown in the problems in the front of this book should not be on the working drawing. If they are needed to obtain the desired points, they should afterwards be removed. They are shown in the problems for explanation, and it is sometimes better to work out the problem on a sheet of paper separate from the working draft, and make a pattern or template of the line obtained and transfer same to the working draft, thereby obviating the erasing of lines and spoiling the surface of the draft.

On a shop working draft the base line of the body is generally used for the center line of the half bottom view, therefore some of the lines of the half bottom view must be drawn on those of the elevation. This is apt to confuse a beginner, and is one of the reasons for our caution against putting on unnecessary lines.

Before one can make a working draft of a body, he must have a general idea in his mind of what he is to draw. It is best to first make a small scale drawing, or put the design on a blackboard, where it can be looked over and changed if desirable. The proportions of a vehicle can be studied better from a blackboard.

Special attention should be given to the full-size draft board and tools for making a draft. Well-seasoned soft pine is the best material for the board, which should be made large enough for the full size of the body. The board should not be less than ⅞" thick and jointed and glued up well. Hardwood cleats should be placed under the board, not fastened directly to it, but so as to allow the boards to expand and contract without warping. Have all edges planed off square. This is very important to insure accurate work. If the draft board is made so that it can tilt at any angle, it is more convenient.

The best of drawing instruments, squares, triangles and curves are necessary. It is also desirable to have a bench and vise at hand, so that the different patterns or sweeps may be made as desired. In making these patterns or sweeps, great care must be taken to get them smooth and true, using the eye to ascertain, by sighting along the edge of the sweep.

It is customary to lay out the near side of the body or the left side elevation first. Outline the body in all views as much as possible, and then fill in the detail and construction. Always aim to keep the body as light in weight as possible, consistent with strength, as unnecessary weight in the construction of any vehicle is undesirable.

Provisions should be made to accommodate locks, hinges and other hardware. The ironing and trimming should also be considered.

And so we could go on indefinitely giving advice, but most of these things must be learned by experience. So we advise the beginner to "stick to it" and put forth his best effort in all things. Perseverance and careful work will lead to success.

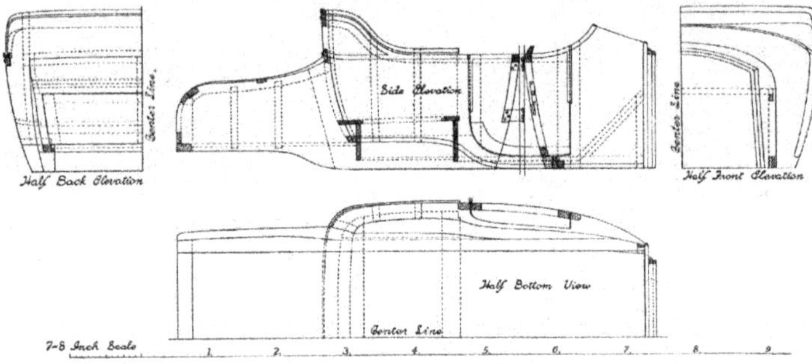

WORKING DRAFT OF TORPEDO ROADSTER.

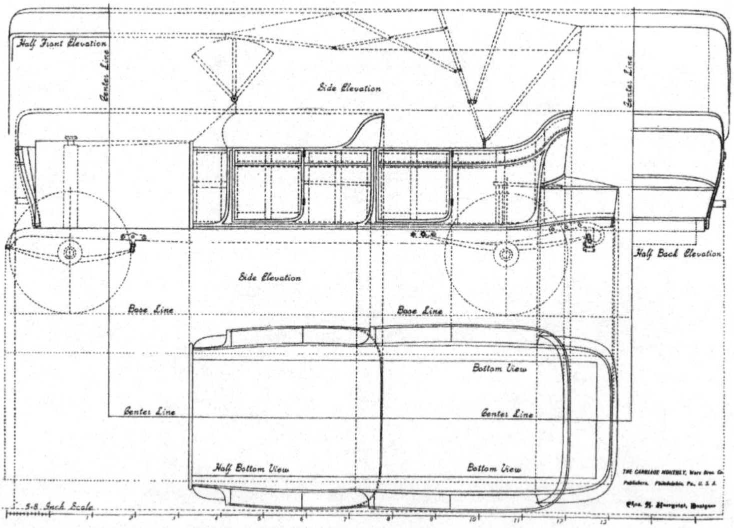

NEW DESIGN OF TORPEDO BODY, WITH CONVEX-CONCAVE TURN-UNDER.

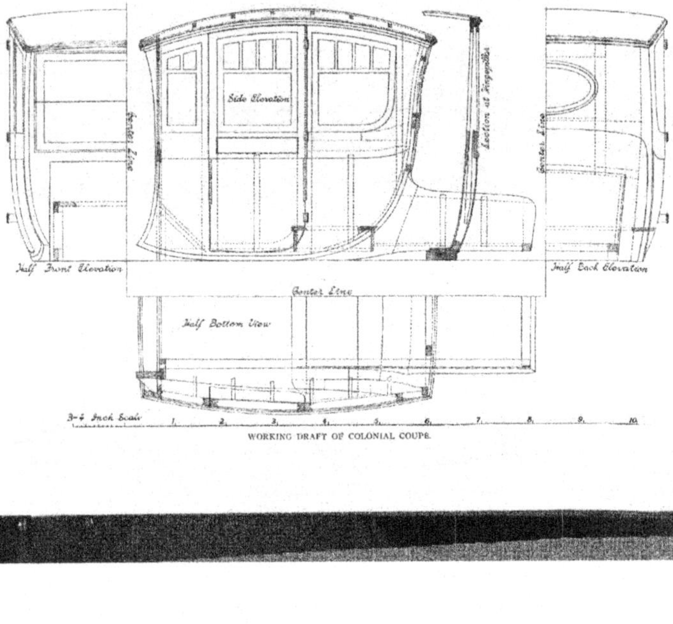

WORKING DRAFT OF COLONIAL COUPE.

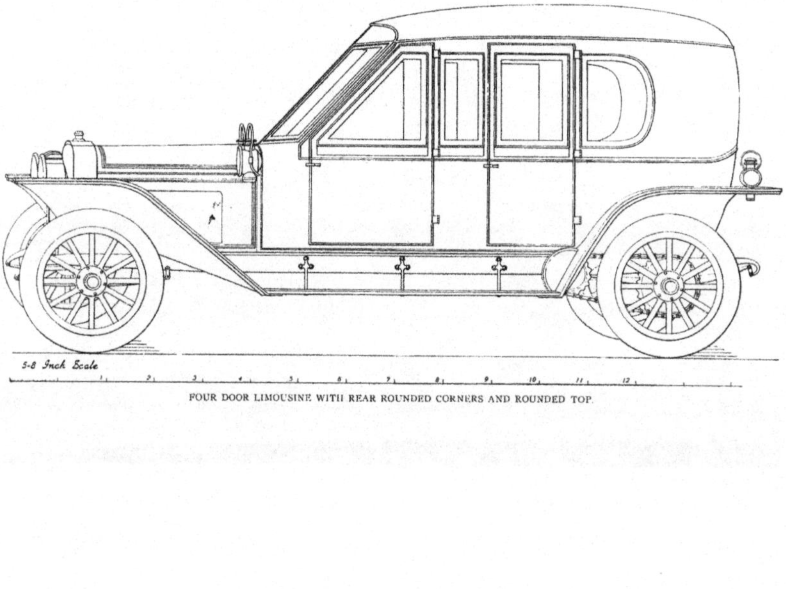

FOUR DOOR LIMOUSINE WITH REAR ROUNDED CORNERS AND ROUNDED TOP.

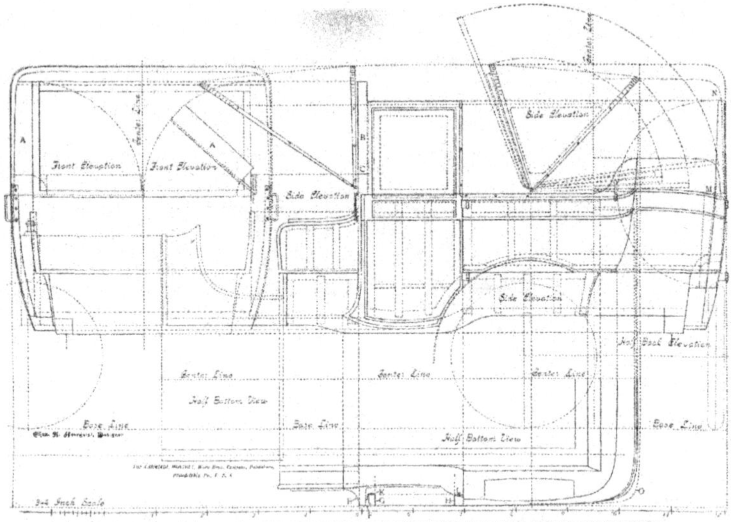

WORKING DRAFT OF TOURING CAR BODY, WITH NEW FOLDING TOP, DROP CHASSIS, ILLUSTRATING FRONT, SIDE AND BACK ELEVATION AND BOTTOM VIEW.

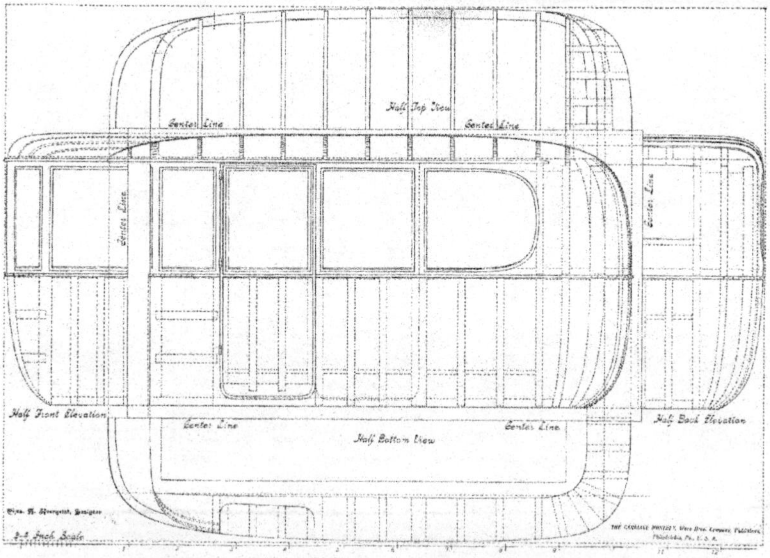

WORKING DRAFT OF LIMOUSINE BODY WITH ROUNDED ENDS AND SWELL SIDE SURFACES.

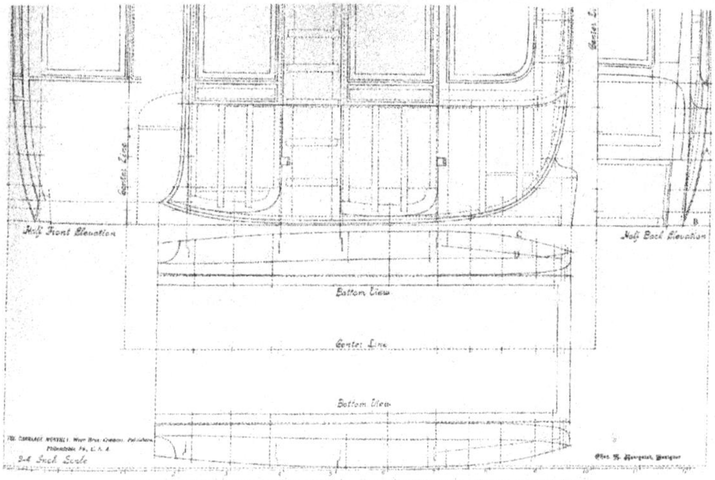

FOUR DOOR CLOSED LIMOUSINE BODY WITH REAR SQUARE CORNERS, FITTED TO A STRAIGHT FRAME.

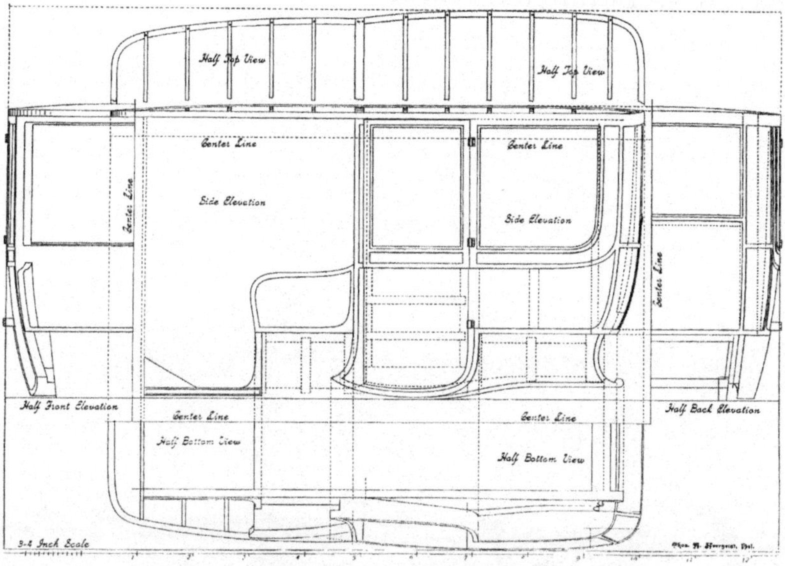

MEDIUM SIZE LIMOUSINE BODY WITH SOLID REAR CORNERS, FITTED TO A THREE-INCH DROP FRAME.

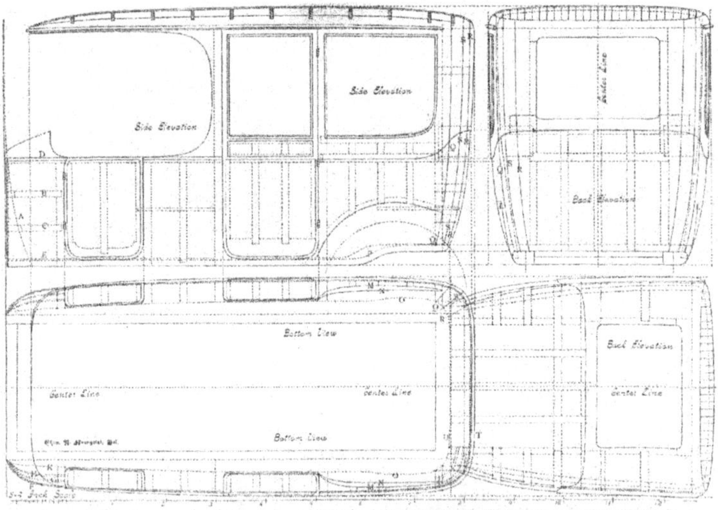

TWO LOW AND TWO HIGH DOOR LIMOUSINE BODY WITH SOLID ROUND CORNERS AND ROUNDED TOP, FITTED TO A THREE-INCH DROP FRAME.

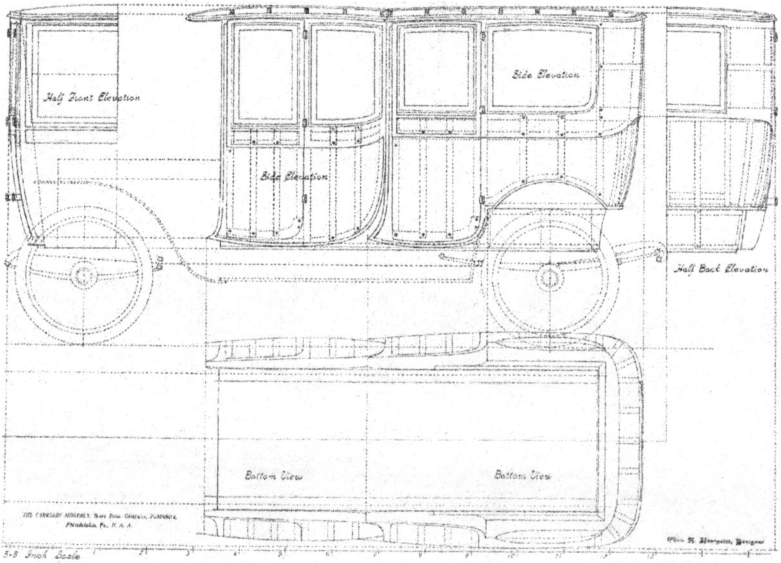

WORKING DRAFT OF AN EXTENSION-FRONT BROUGHAM BODY, SHOWING SIDE ELEVATION, HALF FRONT AND HALF BACK ELEVATION AND BOTTOM VIEW.

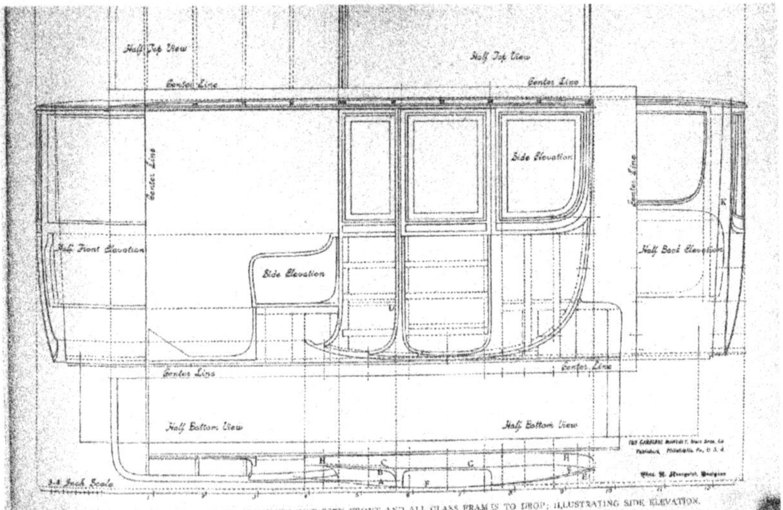

WORKING DRAFT OF LIMOUSINE BODY, WITH ENCLOSED FRONT AND ALL GLASS FRAMES TO DROP; ILLUSTRATING SIDE ELEVATION, HALF FRONT AND HALF BACK ELEVATION AND BOTTOM VIEW.

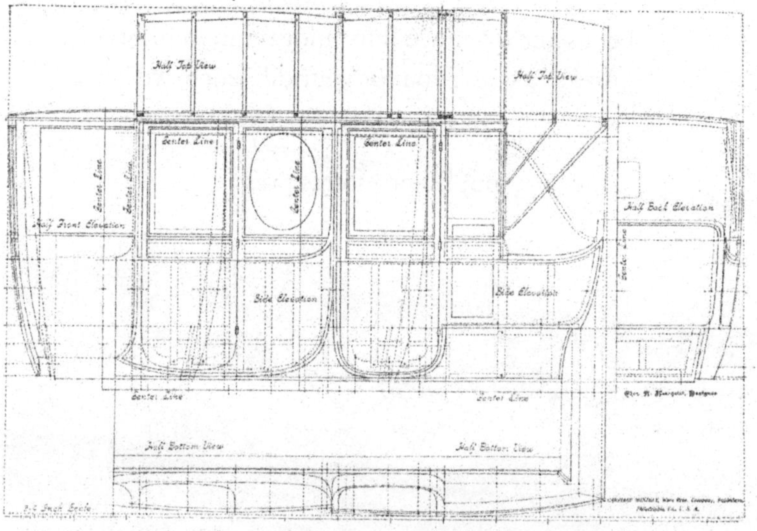

WORKING DRAFT OF LIMOUSINE-LANDAULET BODY, WITH ENCLOSED FRONT AND ALL GLASS FRAMES TO DROP; ILLUSTRATING HALF TOP AND BOTTOM VIEW, HALF FRONT, HALF REAR AND SIDE ELEVATION.
BODY BELOW SIDE QUARTERS IS RECESSED.

Spotted something that isn't quite right?

Let us know and we'll fix it for future reprints.
And as a thank you for your diligence, we'll
send you a free Sothis book of your choice.

publisher@sothis.press

This page intentionally left blank

Except of course for that line above. And this one.
And whilst we have your attention, perhaps you'd
like to follow us on Instagram @sothis.press. Go on.

Made in the USA
Monee, IL
08 November 2026

68106065R00052